Stinking Stones and Rocks of Gold

New Perspectives on the History of the South

UNIVERSITY PRESS OF FLORIDA

Florida A&M University, Tallahassee
Florida Atlantic University, Boca Raton
Florida Gulf Coast University, Ft. Myers
Florida International University, Miami
Florida State University, Tallahassee
New College of Florida, Sarasota
University of Central Florida, Orlando
University of Florida, Gainesville
University of North Florida, Jacksonville
University of South Florida, Tampa
University of West Florida, Pensacola

STINKING STONES AND ROCKS OF GOLD

Phosphate, Fertilizer, and Industrialization in Postbellum South Carolina

SHEPHERD W. MCKINLEY

University Press of Florida
Gainesville · Tallahassee · Tampa · Boca Raton
Pensacola · Orlando · Miami · Jacksonville · Ft. Myers · Sarasota

Printed in the United States of America on acid-free paper

This book may be available in an electronic edition.

22 21 20 19 18 17 6 5 4 3 2 1

First cloth printing, 2014
First paperback printing, 2017

Library of Congress Cataloging-in-Publication Data
McKinley, Shepherd W., author.
Stinking stones and rocks of gold : phosphate, fertilizer, and industrialization in Postbellum South Carolina / Shepherd W. McKinley.
pages cm — (New perspectives on the history of the South)
Includes bibliographical references and index.
ISBN 978-0-8130-4924-3 (cloth)
ISBN 978-0-8130-6461-1 (pbk.)
1. Phosphate industry—South Carolina—History. 2. Phosphates—South Carolina—History. 3. Fertilizers—South Carolina. 4. Industries—South Carolina—History. 5. South Carolina—Economic conditions. I. Title. II. Series: New perspectives on the history of the South.
HD9585.P483U65 2014
338.2'7640975709034—dc23
2013038955

The University Press of Florida is the scholarly publishing agency for the State University System of Florida, comprising Florida A&M University, Florida Atlantic University, Florida Gulf Coast University, Florida International University, Florida State University, New College of Florida, University of Central Florida, University of Florida, University of North Florida, University of South Florida, and University of West Florida.

University Press of Florida
15 Northwest 15th Street
Gainesville, FL 32611-2079
http://upress.ufl.edu

For Richmond Bowens and Cynthia Risser McKinley

Contents

Illustrations

Maps

Figures

Acknowledgments

This journey began in July 1996 at Drayton Hall with a conversation with Richmond Bowens—historical interpreter, son and nephew of phosphate miners, and grandson of slaves. As a tourist, I was spellbound as he described the important industry that began nearby and showed me pictures from his scrapbook. I returned one year later to interview Mr. Bowens and found him to be not only a rich historical source but also an admirable and likeable man. Richmond Bowens died in September 1998 and became the last of his ancestors to be buried in the African American cemetery at Drayton Hall.

I received help, encouragement, advice, and tours from many people throughout the long life of this project. Al Sanders (Charleston Museum) was extremely hospitable as I mined the Edward Willis scrapbooks and other sources. Tracy Hayes and George McDaniel (Drayton Hall), Barbara Doyle (Middleton Place), Ralph Bailey (Brockington and Associates), Mary Miller (Charleston County Public Library), and Ethel Trenholm Seabrook Nepveux and Tom Fetters were enormously helpful in explaining the industry and sharing sources. Michael Trinkley (Chicora Foundation), Claude Thomas (U.S. Vegetable Laboratory), Jim Woodle (Middleton Place), and Charlie Philips (Brockington and Associates) narrated extended trips through remote phosphate-mining areas. Renee Marshall (Huguenot Society), Jim Hare (Ashley River Conservation Coalition), Lynn Harris (South Carolina Institute of Archaeology and Anthropology), Bo Petersen and Arlie Porter (*The Post and Courier*), Jim Rushing (Clemson Experimental Station), George Dent (Southern Dredging), and Peter Wilkerson (South Carolina Historical Society) were also helpful.

Julie Powers Bradley, Vicky Memminger LaRiccia, Lucie Adger, Ann Lesesne Ackerman, Debra Ruehlman, George Smith Adams, Mary Pinckney deMerell Brady, Mary Pinckney Powell, Richard Fitzgerald, Herman Schulte, Paula L. Gibson, Daniel Lesesne, and Claude E. McLeod Jr. all shared information on relatives and properties. Jacque Brotherton (Swift), Bill Tharpe (Southern Company), and Paul D. Ledvina (Mobil) helped me search corporate archives. Barbara Lisenby and Ann Davis (University of North Carolina at Charlotte), Chuck Lesser (South Carolina Archives), and Grace Morris Cordial (Beaufort County Library) aided my never-ending search for sources. Sam Dennis, Don Colquhoun, Tom Downey, Stephen Taylor, Maggy Shannon, Steve Kantrowitz, and Don H. Doyle offered information related to their own work. Pete Daniel, David Carlton, and Phil Scranton provided valuable comments for conference papers. Helpful and patient, Meredith M. Babb and Sian Hunter (UPF) made publishing a smooth process.

I benefited from several mentors and friends whose guidance was crucial to my success in this endeavor. Steve Usselman was an important teacher and adviser, and he pointed the way to southern industrialization. Peter Kolchin carefully and rigorously shepherded this project in its early stages while Sidney Bland and Charles J. Holden offered constructive criticism later. Mentor and friend John David Smith helped to resurrect this project as well as my confidence, and for both of those gifts I am very grateful.

My parents, William R. and Mary Frances W. McKinley, and my brother, Bill McKinley, encouraged me way beyond the call of duty and have always served as role models. The hospitality of my in-laws, James Risser and Virginia and Michael Logan, made much of my research possible. My children, Cameron, Alec, and Gracie, were far too patient. And from that first day at Drayton Hall through the many revisions, my wife Cyndy has unflinchingly and substantially helped me during each stage. Although any mistakes in this work are mine only, she was almost, but not quite, my coauthor.

Introduction

In the eighteenth century, Charleston, South Carolina, was the fourth-largest city in United States, and its planters comprised the richest families in America. The economy was geared toward agricultural export, with most commercial activities dominated by the planter elite—planters, factors, merchants, and lawyers—and subservient to the lucrative shipping of rice and indigo overseas. During the antebellum period, Charleston lost its position as one of the world's premier rice suppliers, and the city declined relative to northeastern cities.[1] As New York, Philadelphia, and other northern ports industrialized, Charleston clung to its more traditional economy. Although Charleston faded as a national economic leader, it emerged as a center of secessionist enthusiasm, eventually hosting the signing of South Carolina's Ordinance of Secession and the opening shots of the Civil War.[2] With little industrialization, a large slave population, and strong regional loyalty, antebellum Charleston and the lowcountry were synonymous with the "Old South." Indeed, white Charlestonians appeared to be fighting change as vigorously as Yankees during the war. By 1865, industrialization in Charleston seemed to be a concept as incongruous as slavery in Boston.

And yet, shortly after the war's end, Charleston's planter elites and former slaves introduced industrialization to the local economy and population. Following scientific advancements throughout the nineteenth century that identified phosphate rock as a primary ingredient in modern commercial fertilizers, lowcountry entrepreneurs and workers created land-mining companies to make use of what they had previously labeled nuisance rocks in local fields. Others established fertilizer-manufacturing factories in the city, on the Charleston Neck, and in other lowcountry locations to make use of the local rock. Individuals started to mine the rivers a few years later, and businessmen and politicians scrambled to secure territories and build

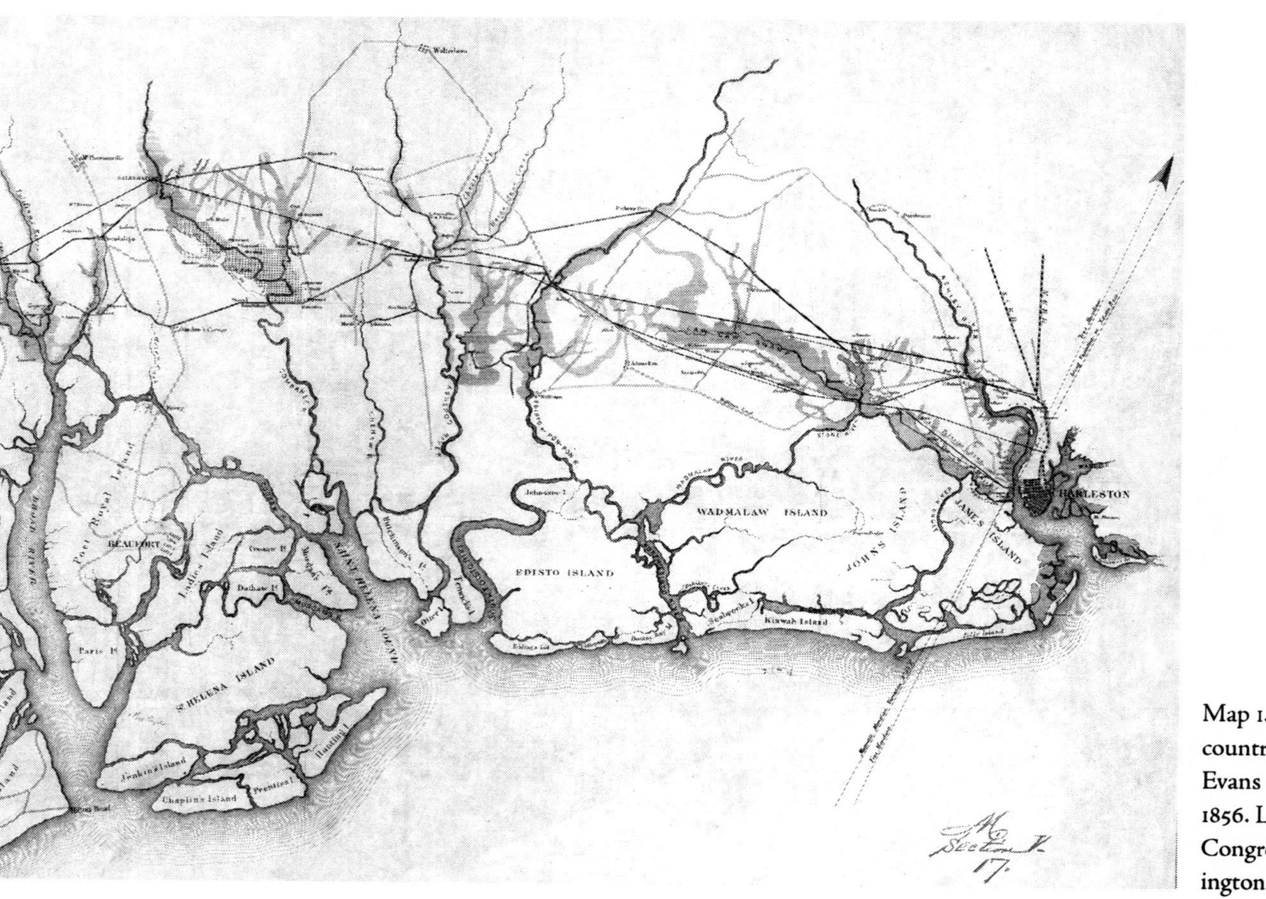

Map 1. The low-country. Walker, Evans & Co., 1856. Library of Congress, Washington, D.C.

companies. Beginning in Charleston and spreading southward to Beaufort's phosphate-rich rivers, factories, railroad spurs, and industrial complexes replaced lightly used areas on the outskirts of the two cities as land mining, river mining, and fertilizer manufacture marked the beginning of industrialization in the lowcountry.

Integrally tied to the antebellum agricultural economy, planter elites recruited northern and southern investors and initiated the three industries between 1867 and 1870. They hired a workforce largely composed of former slaves from lowcountry South Carolina but could not control them. The ex-bondsmen chose independence and mobility over what white managers initially expected: slave-like deference and adherence to the Protestant work ethic. In doing the hard work in the mines, rivers, and factories, freedpeople secured the autonomy born in their particular slave labor system and fashioned lives that matched their conceptions of freedom as well as the political and economic environment.

The following is a narrative case study of the beginnings of South Carolina's three phosphate-related industries—land mining, river mining, and fertilizer manufacturing—that seeks to understand the actions and motives of the primary actors and the process of southern industrialization itself. Workers, entrepreneurs, and managers made important contributions in determining the lowcountry's path to industrialization. This top-down and bottom-up investigation includes elements of business, economic, labor, local, political, and social history and focuses on the intersections of race, class, and political economy in the two decades after the war. Context is crucial. Emancipation, Reconstruction, Redemption, capital scarcity, the legacies of slavery, geography, and a distinct business culture made this version of industrialization unique.

Southern industrialization, different from the "northern model," has yet to be fully understood. This inquiry lies between historical research on antebellum industry and research on postbellum cotton mills, and it builds upon previous scholarship by demonstrating that the phosphate discovery was a significant event because of its timing, participants, location, and impact. Largely ignored by historians, the birth and development of the lowcountry's land-mining, river-mining, and fertilizer-manufacturing industries also suggest the existence of other understudied southern industries emerging in the shadow of the Civil War.

In examining the rise of the three industries, this work contributes several findings applicable to the study of southern industrialization. First, signs of lasting industrialization began to appear immediately after the war in economically traditional Charleston. The industries converged to form the core of modern industrialization in the plantation-dominated lowcountry. Second, southerners and northerners competed in building the three industries during the 1867–84 period. Hardly pawns within a colonial economy, Charleston's conservatives energetically sought to dominate the industries and maneuvered to secure their gains in the statehouse. Moreover, planter elites and internationally prominent gentlemen-scientists eagerly transformed themselves into industrialists after the war. They did not hesitate in pursuing an industrial future. Next, laborers mattered. While elites directed the companies and dominated southern industrialization, freedpeople forced them to negotiate the terms of work. Additionally, freedpeople used phosphate land and river mining to complement their other economic activities, including farming, hunting, fishing, and foraging, and in doing so, formed a shadowy but vibrant economic community. Furthermore, due to the inadequacies of the census and the nature of the work and businesses, the three industries were virtually invisible in the historical record but extremely important to South Carolina's economy. Finally, rock mined and fertilizer manufactured in the lowcountry contributed to fundamental changes in southern agriculture and the American fertilizer industry.

The sum of the findings indicates that historians have only begun to understand the impact of emancipation on southern industrialization. In the lowcountry, elite whites were in some ways a modernizing force and freed blacks were keepers of tradition. In adapting to the significantly different postwar environment, planter elites turned industrialists appeared to be as "freed" from slavery as were their former chattels. No longer "chained" to the slave economy, planters, cotton factors, merchants, and lawyers aggressively, if desperately, pursued new opportunities in the three industries by altering the local environment, converting southern farmers into commercial fertilizer consumers, and expanding antebellum trade relationships with northern and European businessmen. Ironically, it was the lowcountry freedpeople who often obstructed change after emancipation. They sought to protect many of their "freedoms" gained under slavery, including a system of loose supervision and work autonomy. Black miners and factory workers rarely resorted to strikes, instead employing the more subtle negotiations that had been successful under slavery. Featuring surprising elements of continuity and change, the southern economy in the last third of the nineteenth century was a significantly more complex story than that of sharecroppers, cotton mill workers, and middle-class businessmen within a colonial economy. The rise of land mining, river mining, and fertilizer manufacturing in the lowcountry suggests that similarly misunderstood industries and their participants existed throughout the South, and when rediscovered, will contribute to a more sophisticated view of southern industrialization, the postwar southern economy, and the impact of emancipation.

In focusing on the planter elite and freedpeople, *Stinking Stones and Rocks of Gold* supports recent findings in the classic, albeit dated, debate between James Oakes, who described antebellum planters as entrepreneurial, and Eugene D. Genovese, who claimed that they were "prebourgeois" and impeded industrialization. Many historians now argue that the antebellum southern economy and its leaders' business practices can be best described as "modern" and capitalistic. In the lowcountry's three phosphate-related industries, the transformation from antebellum planter elites to postbellum industrial entrepreneurs appears neither slow nor half-hearted. This indicates that these men were not anti-industrial before the war.[3] Part of a dynamic commercial environment linked to international markets, lowcountry planters, factors, lawyers, and merchants nonetheless were constrained by the cornerstone of their economy, slavery. Emancipation resolved the contradictions and helped unleash industrialization in the South.

While modern in their entrepreneurial pursuits, lowcountry postbellum business elites built upon a conservative ideology formed before the war. Genovese's description of slaveholder paternalism as the belief in a "stable, hierarchical, . . . consciously elitist" social order illuminates the foundations of the ideology of the businessmen who controlled the early land-mining, river-mining, and fertilizer-manufacturing companies. Fred Bateman and Thomas Weiss add that planters were conservative rather than anti-industrial in their investments. This analysis of the three industries highlights an elite that altered and expanded its conservative ideology after 1865 in the economic, social, and political arenas. Lowcountry elites continued to assume their leadership roles as natural.[4] Just as planter paternalism contained elements of hierarchy and control, so too did postwar conservatism, in this case, over the new industries and the developing social order. Keeping a lower profile during years of political chaos, these conservatives spoke through business-friendly newspaper articles and industry pamphlets, quietly asserting their claims to power and control.

This research refines previous analyses of South Carolina's conservatives. William J. Cooper Jr.'s portrayal of a vague and nostalgic Conservative (elite Democratic) ideology fails to fully capture the beliefs of those lowcountry conservatives who participated in the three industries.[5] While hardly repudiating the seemingly glorious past, planter elites anxiously but eagerly transformed themselves into industry insiders and appeared to be more men of the New South than of the Old. Adapting the three ideological pillars of race, class, and history to postwar realities, conservatives accepted the Industrial Revolution and the New South while, in Charles J. Holden's words, "skillfully . . . retaining the sense of hierarchy, deference, and sentimentality" of the Old South. Contrary to Don H. Doyle's more negative interpretation, lowcountry conservatives were actually dynamic postwar actors who recognized the inevitability of change and resolved to direct it.[6]

Exposing elements of continuity and change within the lowcountry's industrialization, this study complicates an important debate in southern historiography. Although planter elites, more than "new men" from the middle class, created the conservative modernization in and near Charleston, the result was the start of vigorous industries and permanent industrialization. Contrary to C. Vann Woodward's more general assertions, early industrialization in Charleston did not lead to a "colonial" relationship with the North. Although much of the phosphate rock went to European and northeastern cities, lowcountry industrialists began in the late 1860s to establish their own fertilizer-manufacturing companies to use the locally mined rock. Not until the late 1890s would Charleston businessmen lose control of their companies and become colonial suppliers. But by then they had helped shift the center of the American fertilizer industry to the South.[7]

Much of the historical literature separates southern labor into agricultural and industrial categories and chronicles the struggles of blacks and whites on farms and white workers in cotton mills. Douglas Flamming and David Carlton include a third category, "urbanization," and recognize and pursue the interrelatedness of all three. The land-mining, river-mining, and fertilizer-manufacturing examples suggest that more research needs to be done into a regional economy—invisible to labor organizers, politicians, and the census—with black workers moving fluidly among field, factory, and city. Land- and river-phosphate mining shared the "invisibility" Gavin Wright attributes to the lumber industry, because they were extractive industries characterized by temporary bases and dominated by black workers. Fertilizer workers in Charleston and Beaufort, a significant part of southern industrialization, also escaped the notice of census takers, who were unsure how to categorize them, and of southern labor historians, who were searching for aborted unionization.[8] The failure of the 1870 and 1880 censuses to accurately portray the work habits of lowcountry blacks suggests that historians have made similar oversights elsewhere in the South. What of the workers who spent part of each year in the factory, on the farm, and in the mine?

The three phosphate-related industries demonstrate that a significant number of lowcountry freedpeople were not trapped in the exploitative postwar agricultural labor system and suggest that many blacks throughout the South may have pursued industrial alternatives as well.[9] Certainly the region's economy was not healthy between the Civil War and World War Two, and most African Americans had the worst jobs and living standards. However, evidence exists of a quiet economy—including fishing, hunting, an internal economy, and temporary jobs—that enabled black families and workers to survive alongside, and only occasionally within, the inhospitable white economy. Phosphate land and river miners, and to a lesser extent fertilizer workers, passed like shadows across the historical scene, partly because they did not want to be detected.

Historians of the postbellum South are getting better at discerning the economic footprints of populations that leave few records. This examination detects part of a mobile labor pool, but other studies of southern labor have begun to raise questions concerning the temporary and permanent transitions of black workers from rural to urban and agricultural to industrial settings. For example, do various behaviors within the black working class indicate alternative values? Was what appeared to whites as black resistance to free-labor assumptions—economic rationality, self-discipline, and market responsiveness—the elaboration of prioritized values such as family, education, and autonomy?[10] As this investigation suggests, workers behaved and lived in ways that made sense to them and did not necessarily conform to the prevailing logic of contemporaries or historians. Oppression breeds a world of hidden lives, not just isolated acts, and southern historians need to pry open that concealed world.

The world of lowcountry black workers was but one example of the nation's adjustment to the implications of freedom and the ensuing concern over social order after the Civil War. Freedom, free labor, and independence evolved from antislavery rhetoric to national values after emancipation, but definitions of each lacked consensus. Should freedpeople be free to change employment or livelihood every day, or would this type of economic activity endanger the social order? Amy Dru Stanley and Heather Cox Richardson have begun to measure the ideological foundations, historical backgrounds, and practical implications of nonstandard economic activity in the decades after the war.[11] While northern workers responded to class exploitation with growing militancy, southern black workers responded to racial and class-based roadblocks by increasingly withdrawing to a hidden world.

This inquiry continues a trend toward a closer examination of black southern workers, but it also represents an attempt to integrate analyses of those workers experiencing industrialization, agricultural change, and urbanization. Mark V. Wetherington's investigation of Wiregrass Georgia, Julie Saville's work on rural South Carolina, and Bernard E. Powers Jr.'s analysis of black Charleston are excellent histories that nevertheless are less sensitive to black workers whose working and personal lives straddled the city and the country.[12] Further investigation may uncover early indications of what became apparent decades later in other areas of the South—a form of modernization within a labor-intensive, extractive economy in which agriculture became a part-time occupation.

Although they are significant examples of southern industrialization, phosphate land and river mining and fertilizer manufacture did not revolutionize every aspect of the lowcountry's economy. Even at the three industries' peaks, they did not provide sufficient economic linkages to revitalize an economy that had been in decline for much of the nineteenth century, nor did they alter the region's long-term structural problems or substantially raise the population's low standard of living. By 1900, the lowcountry was one of the poorest places in the United States, and the chasm between the two races was as wide there as it was throughout the rest of the South.[13] This examination ends in 1884, near the height of both mining industries and when the fertilizer industry was healthy. By the 1920s, land and river mining had disappeared, but Charleston's fertilizer mills were thriving. Southern but not local businessmen had consolidated the fertilizer companies, and the city's industrial district (founded by the fertilizer industry) had attracted many other companies. Fertilizer manufacturing did not make all Charlestonians wealthy, but it constituted the vanguard of industrial modernization in the lowcountry during the century after the Civil War.

In the first modern scholarly work on the three industries, Don H. Doyle and Tom W. Shick refer to South Carolina's phosphate "boom" as an "industrial spasm" that proved to be a mere "stillbirth" of the New South.[14] By emphasizing the eventual decline of the land- and river-mining industries, the two historians underestimate the magnitude of the transformations—of planter elites, freedpeople, and fertilizer consumers—already apparent by 1884. *Stinking Stones and Rocks of Gold* focuses on the first decades of the three industries, with an eye to the beginnings of permanent change in the lowcountry and beyond.

1

Antecedents, Precedents, and Continuities, 1800–1865

The rocks seemed to be everywhere, but no one knew their value. Francis S. Holmes was a nineteenth-century planter, slave owner, and gentleman-scientist living next to the Ashley River, northwest of Charleston, who followed local tradition and directed slaves to remove the "useless nodules" from his fields. Before the discovery that phosphate-based fertilizer could reinvigorate worn-out fields, planters considered the phosphate rocks physical obstacles to agricultural production.[1] Industry chronicler Edward Willis testified that most local plantations contained piles of the seemingly worthless rocks and that specimens weighing up to several hundred pounds had been found on or near the surface. Travelers on the Dorchester Road labeled the rocks "stinking stones," because the rocks emitted a "fetid" odor when broken. Colonial and antebellum South Carolinians found phosphate rocks on both sides of the Ashley but rarely east of the Cooper River. More rocks lay just below the surface. Later, entrepreneurs and scientists discovered the rocks in rivers near Charleston and Beaufort. Three important lowcountry industries—land mining, river mining, and fertilizer manufacture—emerged within a few years. Industry insiders Holmes, Willis, and Nathaniel Pratt marveled that the rocks had appeared as if by holy design to offer "their" state and section "redemption" during their most dire hour of need, Reconstruction.[2] In reality, the revelation had begun decades before.

The gestational period for the three industries began about 1800 and ended in 1865. Developments within the American fertilizer industry and lowcountry scientific community during the antebellum era established the foundation for commercial exploitation of lowcountry land and river rock after the Civil War. With fertilizer, farmers realized the needs, scientists refined the formulas, and entrepreneurs marketed the new products. The lowcountry, and especially Charleston, was home to a strong tradition of agricultural science before the war, but slavery slowed the adoption of fertilizer and the development of a local fertilizer industry. The slaves themselves formed the most significant feature of the area's society and economy but only tangentially contributed to the discovery of phosphate rock as a fertilizing material. Involving local entrepreneurial and scientific talents, the Civil War was both a catalyst and an obstacle to the development of the industries. Emancipation became the major event in unleashing lowcountry entrepreneurs, scientists, and free laborers to exploit the land and river rock and to develop the region's fertilizer industry.

The American Fertilizer Industry

South Carolina's land-mining companies had their roots in the development of the American fertilizer industry. The national industry began as crude domestic soil enrichment and waste disposal, and evolved, by midcentury, into the mass marketing and production of commercial fertilizers—sophisticated chemical mixtures often made from imported materials and specifically designed to improve soil and yield. Due to increasing soil exhaustion, land scarcity, and market demand, northeastern farmers began to use homemade fertilizers in the first decades of the nineteenth century. Mid-Atlantic farmers, especially those in eastern Maryland and Virginia, experimented with fertilizers in the 1830s. Farmers and planters in the Southeast began to use some fertilizers in the 1850s, while midwesterners, especially residents of the Ohio Valley, began late in the century. Price differences (based on transportation costs and accessibility), density of cultivation, and social structure were important factors in the timing of each region's adoption of commercial fertilizers.[3]

Northern farmers were quick to realize that agriculture drained soils of crucial nutrients and that their soil needed active maintenance. Initially, farmers collected manure and other waste materials from their own farms and spread them on fields. To meet the demands of the growing population, especially in the Northeast, farmers needed greater waste supplies to increase agricultural production. Historian Richard Wines argues that northeastern farmers, by mixing into their fields various materials from the nearby and burgeoning cities and then selling their produce back to the urban markets, fully embraced a "recycling" mentality by the 1840s. In a significant step in the evolution of commercial fertilizer, the farmers abandoned self-sufficiency for dependency on added nutrients from outside sources.[4]

Wines contends that with the substitution of Peruvian guano for locally obtained urban waste in the 1840s, the commercial fertilizer industry in the United States was born. The industry switched from bulky, locally derived, recycled material to modern commercial fertilizer manufactured from imported and nonrenewable resources. The relatively casual collection of waste materials became a large, highly organized business of importing, manufacturing, developing, marketing, and distributing. Entrepreneurs in Baltimore and other cities imported large amounts of guano from Peruvian islands, the South Pacific, and the Caribbean. Fertilizer was now a manufactured and marketed commodity to be purchased rather than collected.[5]

The American fertilizer industry created demand by training farmers to become Peruvian guano consumers. Northeastern farmers easily adopted guano, because it fit into the existing recycling system as well as their established fertilizing mentality. Guano came from city merchants, improved the soil, and helped increase production for expanding urban markets. In addition, the major guano importers were located in New York and Baltimore, making the substance both accessible and relatively affordable. In the North, the transition from farmers' intuition to "book farming" was not difficult. The result of scientific experiment and commercial marketing, complicated mixtures including guano represented merely differences of degree, rather than substantial changes in a farmer's agricultural routine. Indeed, most American farmers referred to guano as "manure" and to the later commercial fertilizers as "manures" or "guanos."[6]

Guano importation and distribution were major milestones for the development of the modern commercial fertilizer industry. Although "book farming" involved scientists, manufacturers energetically marketed fertilizer, selling brand names for the first time to increasingly demanding consumers. Farmers continually viewed the new and improving fertilizers as a necessity, not a luxury.

Guano was not the ideal raw material for the fertilizer industry. It was costly to transport. Foreign governments began by the 1850s to demand more control over their raw materials, and guano costs rose. Northern farmers had become dependent on commercial fertilizer by this time, and rising guano prices gouged their incomes. Most significantly for price and accessibility, guano became increasingly scarce. As Peruvian sources dried up, high-grade alternative supplies of guano proved difficult to find, control, or transport. Guano imports into the United States peaked in 1856 and steadily declined thereafter. Guano had helped to change northern agriculture, but the end of the guano boom was in sight before the Civil War.[7]

Even at the height of Peruvian guano mania, merchants searched for cheaper and more abundant raw materials. "Superphosphates" were substitutes for Peruvian guano made from bones and from phosphatic guanos that helped pave the way for later adoption of South Carolina's phosphate rock. Recognizing in the 1850s the agricultural value of what came to be called bone phosphate of lime (BPL), American manufacturers burned, pulverized, and dissolved (in sulfuric acid) bones to make a fertilizer material labeled "superphosphate." Manufacturers utilized a similar process on phosphatic guanos from Caribbean and Pacific islands, also calling the result "superphosphate." Although both superphosphates were attempts to supplant Peruvian guano as the favorite fertilizer, they managed to gain only a small share of the fertilizer market. And as was Peruvian guano, bones and phosphatic guanos were expensive, scarce, and of inconsistent quality. In order to keep the retail price competitive with Peruvian guano, producers often added fillers, such as sand, which only intensified farmers' and agriculturists' suspicions.[8]

With an expanding fertilizer industry came better chemistry, or at least more scientific research. Although early superphosphates failed to overtake Peruvian guano, they did help to introduce a change of perception in soil chemistry. Aware that urban waste materials contained low amounts of important nutrients, scientists experimented to find the correct balance of nitrogen, potassium, and phosphorus to reverse soil exhaustion. Adding to the soil potassium from ashes, nitrogen from Peruvian guano, and phosphorus from superphosphates, commercial and academic scientists hoped to improve crop yields and extend soil life. Until the introduction of superphosphates, most scientists followed the "ammoniacal," or "organic," theory in claiming that nitrogen was by far the most important factor in restoring worn-out soils. Phosphatic guano merchants, in aggressively promoting their new product, helped to change scientific orthodoxy by promoting Justus Liebig's "mineral theory." In 1843, the German agricultural chemist argued that phosphates and other minerals were more important to the soil than ammonia content. Adding sulfuric acid to make crushed bones more soluble, Liebig proved that phosphate of lime, not gelatin, was the fertilizing element in bones. By eliminating grease and gelatin from bones, he also proved that phosphate of lime from bones was identical to phosphate rock. While exaggerating nitrogen's faults and mineral phosphates' benefits, the merchants helped popularize the need for phosphate in fertilizer. Although English and American manufacturers had added small amounts of phosphate rock to commercial fertilizers since 1845, the material did not immediately assume major importance. In 1855, John Ketterwell added Mexican guano, which contained no ammonia but high BPL (50–60 percent), to Peruvian guano to produce Ketterwell's Manipulated Guano. In the decade after the Civil War, "complete" fertilizers—those that included high concentrations of potassium, nitrogen, and phosphorus—gradually became the industry standard. Most importantly for the South Carolina fertilizer industry, with the dissemination and acceptance of Liebig's theory, a large and increasing demand for phosphate rock began.[9]

Fertilizer in the South

Although fertilizer consumption increased throughout antebellum America, it took a significantly different path in the South. While northern farmers identified and acted on soil problems, and substantially increased their fertilizer use over the first half of the century, southern planters began talking about soil exhaustion in the 1830s but used comparatively little fertilizer before the Civil War. Scientists and agricultural reformers published papers and gave speeches for agricultural societies, and letters to the editor after 1850 reflected a growing sense of panic about southern soil conditions. This sense of alarm only increased as war and the consequent need for self-sufficiency approached. However, few southern fields received adequate amounts of any fertilizer, and many farmers and planters generally avoided commercial varieties. Use of manure was scattered and ineffective. Peruvian guano and its substitutes made some headway in the upper South and near coastal cities in the 1840s, but dealers' hype obscured the reality that only a small minority did more than experiment with the materials before the late 1850s.[10]

Economically based on plantations and slaves, the antebellum South held fast to soil-depleting agricultural methods and remained a poor candidate for increased commercial fertilizer use. The result was an agricultural system that exacerbated the national fertilizer industry's major weaknesses—price, quality, and supply. While successful northern farmers intensively cultivated smaller fields near cities using family labor, southern planters farmed huge tracts of rural land with slave labor. Planters and yeomen farmers in the South found the new commercial fertilizers too expensive, uneven in quality, and often inaccessible.

Rural southerners did not become enthusiastic fertilizer consumers before the war. Inadequate transportation networks increased fertilizer prices and decreased availability. Poor communication networks left rural farmers ignorant of proper fertilizing methods. Immature channels of capital flow left most without a way to purchase fertilizer on credit. The South had comparatively few urban centers, and consequently, the vast majority of farmers and planters lived far from city waste, the keystone of the North's recycling system. No farm could supply enough manure for adequate fertilization. Without a homegrown fertilizing mentality, southern farmers developed and maintained a severe skepticism toward "book farming" and the industry's Yankee promoters. Most fundamentally, the limitations of slavery and the plantation system made the extensive farming of large tracts a logical option but intensive farming a virtual impossibility. Only the most ardent reformers were willing to undertake a major commitment to manure collection, storage, and distribution. Fertilization, crop rotation, and deep plowing made little sense to planters who controlled an indifferent labor supply. Centuries of this type of farming left southern soils nutritionally deficient. While planter-led southern state governments defended their peculiar institution as the ideal social and agricultural system, the reality was declining crop yields and poor soil conditions. Commercial fertilizers made little headway in this environment.[11]

Historians Eugene Genovese and Gavin Wright suggest several specific obstacles to greater fertilizer use within the slave South. Genovese argues that agricultural reform in the Southeast, including the adoption of fertilizer, proceeded "slowly and painfully" and that the region's failure to fight soil exhaustion became "one of the most serious economic features of its general crisis." The slave labor system contributed to poor agricultural practices, low capital accumulation, and the inability to buy adequate amounts of commercial fertilizers. In a society whose wealth was tied up in human capital, little liquid capital was available for fertilizer purchases. Wealthy planters in the Deep South owned slaves and land but not cash. Only in Maryland and Virginia were planters, through sales of surplus slaves to the Lower South, able to afford improved agricultural methods, including the use of fertilizer. Genovese contends that in passively resisting their condition, slaves were careless with tools and in cultivating fields, and consequently, planters felt that any program of fertilization would be "of doubtful outcome" with slave labor. The vast increase of slave supervision, as more slave labor would be needed to gather, store, and apply manure, discouraged most planters. Even had planters been willing to increase supervision, inadequate manure supplies were another roadblock. Antebellum planters generally kept a small number of livestock, and consequently, manure for cotton fields was in short supply.[12]

The peculiar institution helped to undermine planters' need to practice intensive farming. As Wright suggests, the mobile "laborlords" had two-thirds of their wealth tied up in slaves and were primarily interested in maximizing labor output, not land values. Planters were "land killers" and considered land productivity, town building, internal improvements, and a modern financial system low priorities. Many considered moving to the Southwest a better alternative than improving their "old" soils. Those who stayed in the Southeast kept alive a tradition of indifferent soil husbandry that left only the most successful planters able to afford expensive fertilizers. Not surprisingly, the exhortations of southern agricultural reformers made little headway with laborlord elites. While Wright argues that planter priorities rather than the quality of slave labor limited agricultural reform in the South, both Genovese and Wright think that soil exhaustion was a fundamental weakness in the slave South. And soil exhaustion was an important part of the cycle involving poor yields, lack of capital, expensive slaves, and abundant land that helped discourage planters from purchasing fertilizer.[13] Neither manure nor commercial fertilizer made much of an impact across the South before the Civil War.

The slave system undermined the efforts of agricultural reformers in the South. Antebellum planters' bombastic rhetoric on the national political stage contrasted sharply with their anxious discussions in southern agricultural societies and journals. Agricultural reformers and proslavery ideologues, such as Edmund Ruffin and South Carolina Governor James Henry Hammond, preached improved soil management, especially fertilization, but the response from their fellow planters was underwhelming. Given the costs of manuring, Ruffin and Hammond began to advance the mining and spreading of marl as the answer to soil exhaustion. Beginning in the 1820s and continuing through the 1850s, Ruffin promoted marl as the ideal fertilizer for Virginia and the South. Rich in calcium carbonate, the clay-like substance was abundant and easily accessible in parts of the South, including the South Carolina lowcountry, and was significantly less expensive than Peruvian guano. Hammond solicited Ruffin's 1843 agricultural survey of South Carolina and experimented with and promoted marl throughout the antebellum period. The labor-intensive marling reinforced Ruffin's belief that his reform agenda had to be built on the continued existence of slavery, but poor sales of his published works underscored the contradictions between scientific agriculture and slavery. Labeled "book farmers," Ruffin and Hammond failed to change farming practices in the state and the South.[14]

Despite structural weaknesses in the antebellum southern economy that discouraged the use of commercial fertilizers, northern manufacturers and southern distributors continued to try to open the southern market and establish a "gospel of guano." They faced skeptical and reluctant consumers. Skepticism, however, was only one of many enemies for Peruvian guano sellers. Guano was difficult to obtain in many parts of South Carolina, especially the upstate, due to poor transportation. Across the South, inland and rural farmers had similar difficulties. Where guano was available, high costs and uncertainty over how to apply the substance discouraged many potential users. In 1853 William Allston Gourdin established the first guano dealership in Charleston but paid a carrying charge of nine dollars per ton—an addition of approximately 15–20 percent of the total cost—to import it from Baltimore. Thus, while only wealthy planters from the lowcountry had access to guano, the size of their vast holdings made buying adequate amounts of the material unlikely. Across the South, studies revealed little or no use of manures or commercial fertilizers, despite the "mania" in the North during the 1840s and 1850s. Even prosperity could not produce change. The agricultural boom of the 1850s helped to continue the anti-fertilizer trend in the South. The region's farmers seemed to be technologically complacent in the midst of "dizzying prosperity."[15]

For those who did purchase the substance, fertilizing the fields with Peruvian guano was expensive and time consuming. At ten to fifteen dollars per acre, the sum constituted a major expense for owners of large plantations. Slaveholders worried that costs would rise further if additional labor and supervision were needed to adequately fertilize the large plantations. Others questioned whether slaves could be induced to fertilize the fields properly. While many agriculturists questioned whether guano affected the soil enough to justify the additional effort and high costs, another large faction believed that it was too much of a stimulant, and rather than improving the soil quality, it exhausted the soil at an accelerated rate.[16]

Guano quality also was an important issue. South Carolina and other southern states lacked fertilizer inspectors until after the war, and farmers suspected, often for good reason, that Baltimore's guano merchants adulterated the product with dirt and other fillers. Another quality-related problem came to light in the 1860s as supplies of high-quality Chincha guano dwindled, and American importers switched to lower-grade guanos. Only in the last few years before the war did residents of Georgia and the Carolinas become interested in Peruvian guano.[17]

Bone and phosphatic guano superphosphates, while not transforming fertilizer use in the South, helped to increase consumption by the region's farmers and planters. The South Carolina Agricultural Society's endorsement of Rhodes Super-phosphate in 1860 helped popularize the fertilizer, but its price, substantially less than Peruvian guano, made the real difference. While too expensive for the vast majority, fertilizers were becoming affordable. Still, skeptics urged caution. In December 1859, the editor of *The Farmer and Planter* warned, "the mania . . . for the phosphatic manures seems to be very great. Every paper is full of puffs and advertisements. It is well enough to look into the business."[18] Southern demand for fertilizer was increasing on the eve of the war, but caution still reigned supreme. Some planters and farmers realized the need for fertilizer and accepted its basic virtues, but affordability and accessibility were the most important impediments to its widespread use.

The Gentlemen-Scientists

While southern farmers reluctantly inched their way toward greater demand for commercial fertilizers, gentlemen-scientists in the South Carolina lowcountry made significant contributions to the international study of natural history and agricultural science. Their passion for paleontology, geology, and soil chemistry led directly to the discovery, exploration, and adaptation of the state's phosphate reserves and to the manufacture of phosphate-based fertilizers. War interrupted but did not derail the progress of South Carolina's homegrown scientists, and their scholarship and leadership continued after the war, influencing the postbellum entrepreneurs and their scientific advisers. Gentlemen-scientists were also instrumental in founding the first mining and manufacturing companies, serving as presidents, chemists, or superintendents.

Near Charleston, geology, paleontology, chemistry, and planting came together in a unique community of gentlemen-scientists, who, in the antebellum period, began to lay the groundwork for changing soil chemistry and fertilizer use in the South. The fossil-rich South Carolina lowcountry shared the South's problems with soil exhaustion, and many of its planters and their friends collected fossils and fauna, wrote about soil chemistry, studied geology, and experimented with fertilizers. They also kept abreast of scientific advances and corresponded with internationally known researchers. The result was an unusually rich history of scientific experiments, debates on soil exhaustion, and knowledge of local geology. This scientifically and agriculturally linked environment provided the foundation for the South Carolina phosphate land-mining, river-mining, and fertilizer-manufacturing industries after the Civil War.[19]

Antebellum Charleston's scientific community became, by the 1850s, the Southeast's "center of scientific inquiry." The city featured several intellectual and scientific societies, and several of its scientists achieved international recognition. Natural history was the passion of the lowcountry's gentlemen-scientists, as well as a popular obsession in Europe and the eastern United States between 1810 and 1860. In an age when the line between professional and amateur was blurred, the generalist—often a planter, clergyman, doctor, or politician—wrote natural histories geared toward professionals and lay audiences alike. Natural history also served as a way to understand God's order. Many of South Carolina's naturalists subscribed to natural theology and inherently denied the evolution or extinction of any species. After the 1859 publication of Charles Darwin's *On the Origin of Species*, Holmes and other lowcountry naturalists struggled to reconcile their frequent discoveries of dinosaur bones with their rejection of evolution and extinction. Popular interest in natural history began to decline before 1860, and amateurs like Holmes would find themselves marginalized among increasingly specialized and professionally trained scientists in the coming decades.[20]

Although Holmes and his lowcountry peers would never restore the Charleston scientific community to its 1850s stature, their experiences were not wasted. Antebellum explorations and experiments paved the way for the formation and maturity of the postwar land-mining, river-phosphate mining, and fertilizer-manufacturing industries in South Carolina. Geologists Ruffin and Michael Tuomey systematically explored what became phosphate country. Chemists Nathaniel A. Pratt and Charles U. Shepard Sr. made significant contributions to chemistry before and after the war. Finally, gentlemen-scientists and lowcountry natives Holmes and St. Julien Ravenel built the foundations of lowcountry agricultural science before the war and played integral parts forming and advising companies after the war. Indeed, Holmes became a founding father of the land-mining industry, while Ravenel played a similar role for the fertilizer-manufacturing industry.

Francis Simmons Holmes was a Charleston native and planter, operating the 811-acre "Springfield" plantation owned by his brother-in-law, George Alfred Trenholm, in rural Charleston County. Holmes became an enthusiastic amateur geologist, paleontologist, and scientific agriculturist. With thirty slaves, Holmes, following Ruffin's advice, began experiments fertilizing with marl in 1832. His passion and his hard work with the soil, the subsoil, and the fossils found beneath it led Holmes to become friends with some of the leading naturalists of the antebellum period.[21] In 1837, Holmes discovered on the Ashley River's west bank a "number of rolled or water-worn nodules, of a rocky material filled with the impressions or casts of marine shells." Holmes was holding phosphate rocks. He noticed that the rocks were "scattered" over the land, except for the cultivated fields, from which they had been tossed to the side into piles, so as not to interfere with planting. Five years later, he showed the rocks to Ruffin.[22]

In 1821, the Virginian Ruffin began experimenting with marl to improve soil conditions. In the 1830s, he published *An Essay on Calcareous Manures* and began a monthly agricultural journal, the *Farmer's Register*. In 1842, with Governor Hammond's blessing, Ruffin became the agricultural surveyor of South Carolina. A fellow agricultural reformer, Hammond warned his old friend that South Carolinians "will receive you cordially everywhere, but I cannot promise that they will go far to *meet* you. . . . You must have expected in your task to meet with much indifference, much obstinacy, & some opposition."[23]

Despite Hammond's warnings, Ruffin began his survey with high hopes of converting South Carolina's planters to the gospel of marl and spent "nearly all the time" surveying the lowcountry's abundant marl deposits. Holmes showed him some Ashley River phosphate nodules, and Ruffin later described them as "lumps of stony hardness, full of impressions of shells, containing six per cent of Carbonate of Lime." Based on that analysis, Ruffin dismissed the rocks as "useless as a fertilizing substance." Focusing on carbonate of lime instead of BPL and ever the "prophet of marl," Ruffin was more interested in the river's many outcroppings of marl. He found that great beds of marl underlay much of the lowcountry but was surprised how little South Carolinians knew about what he considered important agricultural matters. His survey was eight weeks old before he witnessed a marling operation, and many planters he met could not identify marl. Completing the survey "greatly disappointed" in the number of true converts, Ruffin published his results late in 1843.[24]

Ruffin's visit inspired Holmes to continue fertilization experiments with several materials, especially marl. As his slaves practiced the "onerous" task of digging, hauling, and spreading the marl, Holmes found that marling improved yields to varying degrees but that accessing the marl presented obstacles—trees, drainage, and overburden—that nearly outweighed the benefits. And he realized that many planters would be reluctant to divert slaves from the fields and crops to the marl pits. Holmes' prewar marling experiences would prove valuable in solving similar problems during the early phosphate era.[25]

In his experiments, Holmes continued to gain an understanding of the connections between geologic phosphate and soil chemistry. Writing years later, Holmes realized that some of his early results were pointing him toward a focus on BPL. Marl mined close to the surface and applied to cotton and corn produced "greater effects" than marl mined below ten feet. Holmes concluded, with clear hindsight in 1870, that the surface marl contained more BPL than the lower samples, which had had their BPL content diluted over eons. Devoted in the 1840s to Ruffin's gospel, Holmes had "not at the time even suspected" the connection to phosphate rock.[26]

Holmes' interest in marl led to a friendship with Michael Tuomey, the geologist hired to continue Ruffin's survey work. Tuomey believed that it was "difficult, if not impossible, to separate an Agricultural from a Geological Survey," and he convinced Governor Hammond to expand the agricultural survey to include his discipline. Tuomey submitted his *Report of the Geological and Agricultural Survey of the State of South Carolina* (*Survey*) in 1844 and published his *Report on the Geology of South Carolina* (*Report*) in 1848. Holmes volunteered his valuable assistance, and Tuomey included Holmes' marling experiments in both publications.[27] Embedded within Holmes' description of marling woes in Tuomey's *Survey* lay frequent mention of phosphate. Under Ruffin's influence, Holmes, while carefully identifying each layer of earth down to his prized marl, characterized the phosphate rocks as nuisances. Ever thorough, Holmes described the rocks as "closely embedded in stiff blue clay" and of "irregular shape, filled with holes and the prints of shells." Holmes was aware that the rocks, rarely larger than a brick and usually the size of an adult's fist, were a distinct substance, mingled with, but not consisting of, marl, green-sand marl, marlstone, clay, coprolites, conglomerates, fossil teeth, or fossil bones.[28]

Tuomey's *Survey* described the mining operation that was marling. Holmes and "two fellows (prime hands)," opened a twenty-foot square marl pit, four and a half feet deep, in five days. Slaves could pump or bail the pit in ten minutes with comparatively little water seepage. As for transportation, the naturalist used two mule carts driven by two men for a distance of 600 yards. Valuable records for the future land-phosphate mining entrepreneur, Holmes' meticulous account of his marling operation included mining yield and a detailed cost summary.[29]

Like Holmes, Tuomey frequently commented on the abundant phosphate rocks on both sides of the Ashley River within a dozen miles of Charleston's city limits. He observed that at Drayton Hall, phosphate rocks "have been gathered from the lawn and thrown into heaps." But like Ruffin, Tuomey emphasized the disappointing carbonate of lime content in the "marl stone" (phosphate rocks). Tuomey was more interested in the marl beds, which, when following the Ashley River from Charleston, began at Bee's Ferry.[30]

By utilizing elements of agricultural chemistry, Tuomey sought to help southern farmers achieve an "enlightened system of agriculture." Reasoning that plants "analyze soils most accurately" and derive their nutrients from the soil, he declared that "if we continue to abstract these matters, by repeated cropping, and without making any return, sterility must be the result." Tuomey argued that agricultural chemists should be called upon to analyze the plants, soil, and crops, and calculate what must be returned to the system to keep it functional.[31]

Tuomey's *Report* featured the work of Charles U. Shepard Sr., an agricultural chemist whose analysis of marl's phosphate content marked a major step in the evolution of the rocks from annoyance to asset. An assistant to Yale's legendary Benjamin Silliman and a published mineralogist and geologist, Shepard held various posts at Yale, Amherst, and the South Carolina Medical College in Charleston. True to his lowcountry peers, Shepard also had an interest in soil analysis and contributed to Ruffin's 1843 report.[32] In Tuomey's *Report*, Shepard began overturning the gospel of marl by moving lowcountry science toward two crucial insights: that the alleged nuisance rocks contained a high percentage of BPL and that BPL, not calcium carbonate, was the more vital ingredient for soil replenishment and plant growth. As did his many predecessors in the 1840s, Shepard analyzed marl instead of phosphate rock. Despite this near miss, Shepard's early analyses contributed to his realization a decade later that the rocks, not the marl, should be the point of focus. In his *Report*, Tuomey credited Shepard with discovering significant traces (6 percent to 15 percent) of BPL in the marl. Shepard analyzed four different-looking "marls," several samples of which came from the future heart of the land-mining industry. In breaking down almost the entire chemical composition of the marls, not just their content of carbonate of lime, Shepard found significant traces of what he later would realize had leached from the phosphate rocks into the marl, BPL.[33]

Shepard's chemical analyses of the marl in areas north and northwest of Charleston, along the Ashley and Cooper Rivers, and the Pon Pon region along the Ashepoo River, laid the foundations for discovery of the value of phosphate rock. At the Clement, Hanckel, Drayton Hall, Gedding, and Harleston plantations along the Ashley, Shepard found surprisingly high amounts of BPL. Likely influenced by Liebig, E. Emmons, a Dr. Vogel, and other international scientists, Shepard's advocacy of BPL in marl was a significant break from Ruffin's focus on marl's carbonate of lime content. Shepard noted that when planters spread "phosphatic marls" (high-BPL marl), "the maturation of the grain is more perfect, the quantity and quality both, being highly promoted." Shepard would not fix his sights on the phosphate rocks for another decade, but his explicit recognition of the benefits of phosphate on plants was a momentous juncture in the marriage between geology and soil chemistry.[34]

Tuomey was reluctant to stray from the gospel of marl, despite Shepard's findings. Tuomey acknowledged that Ashley River marl was not only "the best in the state," rich in carbonate of lime, but also contained what he termed an "exceedingly interesting ingredient," phosphate of lime. He found 4 percent BPL in marl near Bee's Ferry, later to be the epicenter of the land-phosphate mining industry. Tuomey even made a crude analysis of phosphate rock, mistakenly concluding that the nodules had an insignificant 15–16 percent BPL; postwar analyses consistently recorded 50–55 percent. The man whose career benefited greatly from Ruffin's recommendation argued, "[S]till, I apprehend that the carbonate of lime will always prove the constituent of greatest importance, valuable as phosphates are." Ironically, when Tuomey trumpeted the Ashley marl as being "generally accessible and . . . exposed," and able to be "transported at trifling expense," he unwittingly foreshadowed the utility of phosphate rocks.[35]

Meanwhile, Shepard had a chance encounter in 1855 with a substance of remarkably similar chemical composition to South Carolina's phosphate nodules. When he analyzed a shipment of "stony phosphate" from Mong's

Island in the Caribbean, he found that the "Pyroclasite" and the lowcountry phosphate nodules belonged to "the same species," although the local rock was not as rich in BPL. In another description from what was probably the same incident, Shepard analyzed shipwrecked "Sombrero rock guano," found it rich in BPL, and realized for the first time that "guano, or its equivalent as a fertilizer, may be found in hard rock-like masses." This discovery turned his attention to the Ashley phosphate nodules themselves, not merely the BPL in marl.[36]

In an 1859 address to the Medical Association of the State of South Carolina, Shepard was the first lowcountry scientist to declare that phosphate rocks could be much more than a nuisance. He argued for "a careful examination of all our marl beds, with a view to determine which have the most of the precious phosphatic ingredient." Shepard boldly predicted that "as the supply of guanos from abroad fail, we shall be looked to to fill the vacuum their disappearance will occasion; and it would not be strange, if a few years hence, Charleston, besides supplying her own state, should ship more casks of phosphatic stone to the North than she now receives of ordinary lime from that region."[37] Clearly, Shepard had left behind Ruffin's marl and refocused attention on the rocks that cluttered marl beds and littered fields near Charleston.

Also in 1859, Shepard entered into a collaboration to manufacture fertilizer with Lewis M. Hatch, his son Melvin P. Hatch, and the senior Hatch's brother-in-law, T. P. Allen. A Columbian guano dealer in Charleston, Lewis Hatch previously had hired Shepard to inspect each shipment. Having secured Shepard as the new company's chemist, Hatch proposed to manufacture fertilizer mixing the city's "refuse matter" with sulfuric acid and Peruvian guano. With "special partner" Lewis Hatch supplying the capital and "general partner" Melvin Hatch running the day-to-day operations, the new company was, in the elder Hatch's words, "in every way a success." The Hatches initially relied on bones to supply the phosphate content of their fertilizer but only were able to collect a year's supply. Having concluded that phosphate rocks shared many similarities with rich Columbian guano, Shepard proposed that the Hatches consider "the Ashley River marl or rocks." Shepard's suggestion, the first recognition that lowcountry phosphate had value as a commercial fertilizer, spurred the Hatches to investigate the Ashley River marl beds for rock.[38]

The prospect of a seemingly unlimited and cheap raw material obviously appealed to the Hatches. In the spring of 1860, they and Shepard began to look for the "nodules" on Ashley River plantations. That summer, the Hatches sent samples to Shepard at New Haven for chemical analysis. Shepard was not in New Haven when the samples arrived, so the samples were ground and spread into his garden, yielding impressive results. By fall, Shepard, apparently basing his judgment on the rocks' physical appearance and their performance in his garden, told the elder Hatch, "I found it [the Ashley River phosphate rock] far richer than I expected; so rich, that with it we can drive other fertilizers out of this market, and may soon invade foreign markets." Lewis Hatch immediately hired a steam engine "to push this business."[39] Hatch failed to clarify if his firm began to use the area's phosphate rocks to "compete with the world in fertilizers," but South Carolina's secession on December 20th caused Hatch and company to suspend business.

Shepard did not keep the good news to himself or his employers. Sometime in 1860, he entered a partnership with George T. Jackson of Augusta, Georgia, to manufacture fertilizer. Apparently, the two men struck their agreement without first discussing phosphate rock. Soon, however, Jackson realized that the supply of bones in the region was insufficient for their manufacturing needs, and he discussed the matter with Shepard, who told him about "a large deposit of marl on the Ashley River" that would "answer our purposes." Why Shepard had not initiated talks of fertilizer manufacture with this information is not clear. Equally confusing is Shepard's use of the term "marl" when in fact he meant the phosphate rocks within the marl beds. Shepard sent Jackson samples of the rock, not marl, in the spring of 1861. As did the Hatches, Jackson and Shepard suspended plans for manufacturing phosphate fertilizer after secession.[40]

The War

The Civil War was both a catalyst and an obstacle to the development of the land-mining, river-mining, and fertilizer-manufacturing industries. War mobilization brought together the final pieces of the scientific puzzle as well as many key players in the future industries. Gentlemen-scientists working in the Confederate Nitre Bureau honed skills in chemistry, exploration, and mining.[41] Lowcountry shippers, merchants, and cotton factors turned blockade runners gained or maintained foreign trading relationships and perfected transportation networks. Charleston's mechanics and scientists collaborated on marine inventions. But the four-year emergency also deferred development of the three industries. Despite the evolving attitude that phosphate rocks had commercial potential, Charleston's nascent fertilizer enterprises dissolved with the outbreak of war, and scientific experimentation with the rock all but ended.

Charleston's gentlemen-scientists were loyal Confederates. In January 1861, Holmes showed himself to be a firm supporter of secession, declaring in a letter to the Philadelphian Joseph Leidy, "Black Republicanism has driven us to this measure." While assuring Leidy that their friendship was secure, Holmes argued that "there must be a Southern Confederacy" and that "we will never flinch before a 'Lincoln Force.'" Father of eight and forty-six years old, Holmes volunteered for administrative work rather than soldiering and served as chief superintendent of Nitre District 6, South Carolina.[42]

While supervising his nitre works, Holmes struck up a friendship with Nathaniel A. Pratt, a Harvard-educated Georgian who specialized in geology and chemistry. His Confederate superiors transferred Lieutenant-Colonel Pratt to the Augusta office of the Nitre Bureau, where he spent the remainder of the war. Pratt became the chief chemist (and, later, acting chief) at the Augusta laboratory, the experimental center for the Nitre Bureau, and also was the Confederacy's top consultant for nitriaries and mines in South Carolina, Virginia, Georgia, Florida, Tennessee, and Alabama. His extensive travels demonstrated the Nitre Bureau's comprehensive efforts to harness the South's mineral potential for the war effort.[43] His travels also exposed him to what would become the phosphate region of South Carolina.

Pratt visited Charleston twice in 1864 to inspect Holmes' nitre works. During two trips up the Ashley River, Holmes brought Pratt to Bee's Ferry on the west bank of the river to show him marl deposits, fossils, and another substance. In his postwar book, Holmes claimed that he showed Pratt "coprolites" and that Pratt found 15 percent BPL in the fossilized excrement. Holmes thought that Pratt never saw phosphates during the war, but Pratt, in a promotional work written after the war, claimed that the "supposed Coprolites" were phosphate rock. Aware of Shepard's findings (9–10 percent BPL in the Ashley marl compared to 2–3 percent for Georgia marl), Pratt took "various samples" back to his Augusta laboratory but never analyzed them. Impressed by the marl and other substances Holmes had shown him, Pratt resolved to return to the lowcountry after the war to manufacture fertilizer using local resources. At this point, Pratt had not positively identified the commercial value of South Carolina's phosphate, but he had a hunch that the banks of the Ashley River contained superior (in terms of BPL) fertilizing material.[44]

The war also brought gentleman-scientist St. Julien Ravenel into the phosphate circle. The locally renowned Ravenel shared Holmes' and Shepard's passions for scientific discovery, agricultural adaptation, and business opportunities. Ravenel's father was John Ravenel, a prominent planter, leading cotton factor, and railroad entrepreneur. As Edmund Ravenel's nephew, St. Julien Ravenel was also a member of the lowcountry's scientific aristocracy. Educated in Charleston, New Jersey, Philadelphia, and Paris, St. Julien Ravenel was, like Holmes, interested in natural history, a close friend of Louis Agassiz, an expert in drilling wells, and part of the Charleston scientific community. Ravenel performed countless soil experiments throughout his career, seeking to lower fertilizer costs while improving lowcountry crop yields and diversity. He also established the Charleston Agricultural Lime Company at Stony Landing, his plantation on the Cooper River, using marl mined locally as a lime substitute. Ravenel was yet another lowcountry scientist whose antebellum and wartime pursuits prepared him for a postwar career in the fertilizer industry.[45]

Ravenel worked for the Confederacy in Charleston and Columbia. He had known Ruffin since the 1840s, and during the latter's 1861 visit to Charleston, the Virginian visited Ravenel's lime works, which supplied the Confederacy. As he toured Ravenel's operation, where workers burned marl to produce lime, Ruffin wistfully reflected, about lowcountry planters, that "it is astonishing, and would seem incredible, that highly intelligent men, as are many of these proprietors, should not have used this manure, in its crude state as marl, and all over their land." Ruffin undoubtedly approved of Ravenel's efforts to exploit the marl.[46]

While on furlough in Charleston during the war, Ravenel collaborated with yet another fertilizer industry pioneer, David C. Ebaugh, in an

Figure 1. The founders (*clockwise from upper left*): Francis S. Holmes, Nathaniel A. Pratt, Charles U. Shepard Sr., St. Julien Ravenel. Philip E. Chazal, *The Century in Phosphates and Fertilizers* (Charleston: Lucas-Richardson Lithograph & Printing, 1904).

attempt to rid Charleston Harbor of enemy ships. Responding to Confederate incentives, businessman Theodore D. Stoney enlisted the help of Ravenel, who, in turn, contacted Ebaugh, a talented mechanic and the superintendent of the C.S.A. Nitre Works at Stony Landing. The three men formed the Southern Torpedo Company to fight the Union menace with new technology. Ebaugh, with help from Ravenel, built the world's first "torpedoboat," the CSS *David*, which attacked the USS *New Ironsides* in October 1863, disabling the vessel but failing to end the blockade. The Ravenel-Ebaugh partnership did survive, however, into the phosphate era.[47]

Integral to the war effort, two high-profile Confederates and several cotton factors and shipping merchants turned blockade runners used similar talents as entrepreneurs in the postwar land-mining, river-mining, and fertilizer-manufacturing industries. For many, C.S.A. service was a defining moment in their place in the community as well as their lives. More specifically, the men who managed and financed blockade-running firms gained or maintained international business relationships, entrepreneurial experience, and organizational skills that they used after the war in creating the three postwar industries.

Two future lowcountry fertilizer pioneers, C. G. Memminger and George A. Trenholm, held the same post in the Confederate cabinet. A wealthy landowner, lawyer, and slave owner, the German-born Memminger was a leader in the secession movement in the 1850s and a key player in drafting the Ordinance of Secession. In February 1861, President Jefferson Davis appointed Memminger secretary of the treasury. Like Davis, Memminger faced stiff opposition from factions within the Confederate Congress, and the besieged secretary finally resigned in June 1864.[48] Trenholm succeeded his Charleston friend as treasury secretary, serving until the end of the war, but his service to the Confederacy as a financier and blockade runner was more significant than his brief role in Davis' cabinet. As with several other cotton factors and shipping entrepreneurs in Charleston, Trenholm's antebellum career prepared him to pursue patriotism and profit during the war. He was part of a trend in the 1840s, in which lowcountry merchants began to rival their planter neighbors in wealth and status. Charleston enjoyed boom times before the war, leading the South in Atlantic shipping. The city's businessmen established alliances with northern firms and gained contacts in Europe, especially in Liverpool. Access to foreign capital and businessmen allowed Trenholm and his local peers to make Charleston the center of blockade running during the war. Similarly, those northern and foreign contacts became critical after the war for establishing markets for South Carolina phosphate and fertilizer.[49]

Trenholm spent his entire career with the commission and shipping firm of John Fraser and Company. Founded early in the century, the company was antebellum Charleston's dominant cotton factoring firm and one of the East Coast's top importing and exporting companies. When John Fraser died in 1854, Trenholm controlled the interlocking firms of John Fraser and Company (Charleston), Fraser, Trenholm and Company (Liverpool), and Trenholm Brothers (New York). By 1860, Trenholm was "one of the wealthiest and most influential men in the South," owning steamships, railroads, wharves, banks, hotels, cotton presses, plantations, and slaves. Trenholm's copartners and managers included Theodore D. Wagner (Charleston), James T. Welsman (New York), and Charles K. Prioleau (Liverpool), and all three helped to create Trenholm's postwar phosphate and fertilizer empires. Trenholm was active in state politics, served on several corporate boards, and maintained a close relationship with his brother-in-law and fellow southern nationalist, Francis Holmes.[50]

St. Julien Ravenel also had important family connections to the world of trade and shipping. Ravenel & Company, led by his uncle John and his father, William, was a leading antebellum Charleston shipping firm that did business in Russia, Sweden, England, and Rhode Island. After John retired, William continued the business with two of his nephews and also formed Ravenel & Huger.[51] After the war, St. Julien Ravenel's family and business contacts would provide the funding for several of his phosphate and fertilizer companies.

A third powerful firm on Charleston's Cooper River wharves was James Adger and Company. Irish immigrant James Adger established one of Charleston's premier antebellum cotton factorage, commercial exchange, and coastal shipping firms, and his son Robert took over the business in 1858 when the patriarch died. Although they did not own a lot of slaves, the Adgers became part of the lowcountry planter aristocracy. In the late antebellum era, many of Charleston's planters sent their sons to the city for a "countinghouse education," clerking in a mercantile house; the Adgers began with that practical education and were at home among the city's upper crust. The Adgers also maintained personal and commercial friendships with the Brown family and its shipping empire, which included Brown Brothers and Company (New York), Brown, Shepley and Company (Liverpool), and Alexander Brown and Sons (Baltimore). The gateway to British markets, Liverpool became one of the most important ports for the Adger family businesses, including what would become the dominant river-phosphate mining company after the war.[52]

Although New Yorkers dominated the Atlantic-coast shipping trade, Charleston firms entered the business in the 1840s and were solidly established by 1860. The Adger Line, owned by Charlestonians such as the Adgers and Fraser Trenholm and Company, lost the SS *James Adger* to Union seizure in April 1861, but Trenholm, Ravenel, and other local shippers made Charleston "the Confederacy's lifeline of supply" after the war began. About 80 percent of the ships successfully ran the blockade through the city, contributing supplies to "The Cause" and garnering huge profits for the firms. William Ravenel formed the Importing and Exporting Company of South Carolina with Theodore P. Jervey, William Bee, and John Fraser and Company. Other blockade runners included the Steamship Charleston, Palmetto Exporting and Importing, Chicora Importing and Exporting, Charleston Importing and Exporting, and Atlantic Steam Packet companies.[53] After the war, many of the principals in Charleston's blockade-running companies became phosphate and fertilizer entrepreneurs and chose identical names for their new companies.

Trenholm and his three interlocking firms made the biggest impact on the Confederacy's efforts in blockade running and financing. Trenholm's ships protected Charleston and ran the blockade more than sixty times, and Fraser, Trenholm and Company in Liverpool became the South's financial agent in Europe.[54] Trenholm's Liverpool connection was crucial for the firm and the Confederacy, as well as for Charleston's postwar land-mining, river-mining, and fertilizer-manufacturing industries. The city of Liverpool shared strong business and personal connections with the antebellum South, especially Charleston. Initially, the trade of slaves and, later, cotton, united both cities, and firms such as Baring Brothers and Alexander Brown and Sons worked closely with Charleston's cotton factors. When the war came, and despite Queen Victoria's May 1861 Proclamation of Neutrality, Liverpool's business community actively aided the Confederacy, by helping southern agents thwart Britain's neutrality laws, while making a profit. The city became "the hub" of blockade-running operations, and Fraser, Trenholm and Company's office at 10 Rumford Place was, according to a British government document, "in effect the Confederate Embassy in England." Confederate defeat meant bankruptcy for most of the blockade-running firms and years of legal troubles for Trenholm, but the close commercial ties established before and during the war paid dividends in the postbellum phosphate and fertilizer era.[55] Directors and managers of the blockade-running firms leveraged their wartime contacts, capital pools, and organizational skills to create the three postwar industries.

Although the war damaged Charleston's buildings, economy, and traditions, the "late unpleasantness" did not paralyze the region's scientists and entrepreneurs. The antebellum heritage of agricultural science, natural history, and entrepreneurial drive remained intact in the lowcountry and provided a foundation for the postbellum land- and river-mining and fertilizer-manufacturing industries. The need for inexpensive and readily available fertilizer did not disappear either. Most of the postwar industrial pioneers had served the Confederacy in some capacity, and, after weathering the chaotic years of 1865–66, they returned to Charleston to resurrect the city, their fortunes, and their lives. The sons of Charleston and the lowcountry's leading families began to diversify their careers, seeking alternative pursuits off the plantation.

But the South Carolina lowcountry had changed since 1861. Emancipation removed many of the obstacles to greater fertilizer use, and southern farmers began to purchase the material in prodigious quantities. The lowcountry's grand planter families—including the Ravenels, Draytons, Middletons, and Pinckneys—retained most of their status, some of their land, and a bit of their fortune, but none of their slaves. The freedpeople either moved away, reveling in their new mobility, or stayed, to lay claim to opportunities in a region that also had been theirs for generations. Other ex-slaves moved from the interior to the lowcountry. Of those who stayed, few would work for their former masters and almost none would, in the early years after emancipation, work on plantation lands under a gang-labor system.

Former masters and slaves sought to adjust to a world after chattel slavery and military defeat. This fluid period coincided with the emergence, for the first time in the lowcountry, of an industrial enterprise not directly related to the antebellum versions of the plantation, slavery, and King Cotton. And yet, the new land-mining industry would operate on plantation lands, be worked by ex-slaves, and fuel the new King Cotton. A bastion of the Old South and the birthplace of secession, the South Carolina lowcountry would host a convergence of business trends, agricultural change, and scientific progress. In the process, Charleston and the lowcountry would move, slowly and painfully, toward membership in the New South.

2

The Creation of Industry and Hope, 1865–1870

When Francis S. Holmes returned to Charleston in November 1865, he, like many other former slaveholders, had hit bottom. Late in the war, an "incendiary" had burned his office, personal papers, books, and specimens, and over a year later he complained, "[T]he times are hard and the money is tight."[1] His brother-in-law, George A. Trenholm, had an even more difficult transition into the postbellum era. Fleeing Richmond in April 1865, the Confederacy's last treasury secretary arrived in Charleston, where, in a poignant scene for white southerners, black Union soldiers arrested him. Fraser, Trenholm and Company of Liverpool went bankrupt in May 1867, and Trenholm, his partners, and their firms would not be free of government pursuit for six years. The prominent southern magazine *De Bow's Review* described John Fraser and Company's postwar problems as "one of the heaviest disasters that could have befallen" Charleston and its economy.[2] The editor's comments reflected the belief that as the elite went so did the city.

South Carolina's planter aristocracy did not fare much better than Trenholm. Leading planters lost their agricultural labor and largest investment to emancipation. Charleston and Columbia were burnt shells, and Union armies with black troops occupied much of the state. By 1867, Republicans were insinuating themselves into the fabric of the state's political life. For elites, the war, emancipation, and Reconstruction appeared to have destroyed "their" civilization. The lowcountry aristocracy was especially hard hit. Although the area began declining long before 1861, the war dramatically ended Charleston's claims to economic prestige and leadership.[3] While emancipation and surrender meant new opportunities in the postwar South for freedpeople and many common whites, the future did not look bright to the lowcountry's traditional economic elite.

In the war's shadow, Charleston's gentlemen-scientists discovered that the "stinking stones" along the Ashley River could revolutionize fertilizer manufacture. The men convinced local entrepreneurs and northern acquaintances, and together they began to resurrect the local economy through mining phosphate rocks on plantation lands. The discovery became for some South Carolinians, in the words of Nathaniel A. Pratt, "the means of your redemption." Pratt directed his words to well-educated and wealthy whites of the South, but blacks and lower- and middle-class whites shared in the economic impact. He predicted, correctly, that phosphates would create synergy between the state's agricultural economy, elite businessmen, "willing labor," and "hungry poor." Indeed, phosphate-based fertilizer would begin to transform the South's "waste and desolate places" into productive cotton-growing regions.[4]

Northerners and other southerners joined Pratt, Holmes, and Trenholm in developing the phosphate windfall, but the land-mining industry did not begin as a colonial relationship with the North. Some of the gentlemen-scientists and planters invited northern capitalists to jump-start their companies, but others began their own firms with southern capital. Manipulating and summoning southern patriotism, northerners and southerners began to build companies and brand-name recognition. While memories of war remained fresh among southerners, emancipation released local scientists, cotton factors, planters, and shipping merchants from the constraints of the slave economy and opened new possibilities for them, just as it did for their former slaves. Motivated by desperation, patriotism, and greed, the lowcountry's entrepreneurs responded enthusiastically to the opportunities and began to develop a diversified industry that strengthened rather than challenged the region's agricultural economy.

Eureka!

By 1865, scientists, farmers, and manufacturers in the eastern states were beginning to realize two fundamental problems that had been apparent to only a few of their colleagues twenty years before. First, phosphorus was the chemical element and plant nutrient most deficient in the "old" soils of the East. Second, among all fertilizer ingredients, the phosphorus supply chain had the most substantial bottlenecks. Not surprisingly, given Charleston's rich scientific history, a few of the city's gentlemen-scientists and entrepreneurs sought to address the first problem by starting fertilizer factories after the war, but the second problem remained. Fully appreciating the value of the state's phosphate rock deposits, all located in the low-country, was the "real breakthrough" that solved the supply bottleneck. In turn, South Carolina's phosphate changed patterns of fertilizer use in the South, as well as production in the international fertilizer industry.[5]

The founders of the land-mining industry—Pratt, Holmes, and St. Julien Ravenel—emerged from the war as fertilizer entrepreneurs and became regional heroes. Their later bickering about who deserved more credit for founding the industry revealed not merely the worst in the men's egos but the hopes and dreams of an elite society shattered by military defeat and emancipation. Surprisingly, Professor Charles U. Shepard Sr. was not one of the founders, despite his early analyses of phosphate rock. Shepard was unable to revive his fertilizer business with Lewis M. Hatch after the war.[6] Instead, it was Ravenel who started the postbellum fertilizer industry in Charleston.

In November 1866, Ravenel formed a partnership with David C. Ebaugh and local factor William C. Dukes & Company. Relying solely on local investors, the new business started with $100,000 capital. Given the uncertain state of Charleston's economy since the war, the money raised, or pledged, was an impressive sum. For Ravenel, the partnership was the culmination of fifteen years of soil research, including several years manufacturing lime. Ebaugh contributed to the new venture a wealth of business, mechanical, and labor-management skills.[7] In early 1867, Ravenel and partners established the Wando Fertilizer Company's works on the Cooper River with an iron crusher, pulverizer, and mixer. Without a sulfuric acid plant, they had to import the chemical from the North, but with crushing and pulverizing machinery, they designed Wando to be more than another of Ravenel's lime works. Where the Wando partners would acquire the needed phosphatic raw material remained unclear, but as of mid-1867, they did not appear to include nearby land rock in their business strategy.[8]

After the war, Pratt too had plans to start a fertilizer company. He wanted to use the "native resources of the State through the supposed advantages which this city afforded," but it remains unclear whether Pratt originally intended to use local phosphate rock. Employed as chemist to the North Carolina Geological Survey during 1866 and 1867, Pratt searched both Carolinas for fertilizer raw materials and sought investors in Charleston. Apparently not yet sold on land rock, he did, however, spend a considerable amount of time in the lowcountry from May to August 1867 searching for a factory location. Pratt also spoke with Ravenel, who indicated some interest in investing, but Ravenel joined the Dukes in starting Wando instead.[9]

While searching for a suitable raw material, Pratt crossed paths with Ravenel in the summer of 1867. In Pratt's version of the story, Ravenel encouraged him to examine a rock from a local plantation that Ravenel estimated contained 10–15 percent phosphate of lime (BPL). On August 7, 1867, Pratt enjoyed the eureka moment of South Carolina's three phosphate-related industries when he found 34.4 percent BPL in Ravenel's rock, more than double what the two men had expected. The difference in BPL was not only unprecedented in America but of monumental significance for Pratt and Ravenel as well. While they regarded a reading of 10–15 percent BPL as indicative of a mildly attractive fertilizing additive, a rock with 34 percent BPL was a viable primary raw material for fertilizer. Pratt and Ravenel realized that if they could locate enough rock, their source worries were over.[10]

In Ravenel's version of the story of the eureka moment, he alone "became aware of the true character of the nodule" in the "early" summer and immediately initiated a search for phosphate rocks. When Pratt visited Ravenel in August, Ravenel expressed interest in Pratt's manufacturing plan, because Wando had "at hand a large supply of native Phosphate of Lime, which only needs cheap sulphuric acid, manufactured on the spot, to make it of value as a fertilizer." Ravenel gave a sample to Pratt so that the latter could "satisfy himself, *by his own analysis*" of the rock's commercial value. Although both understood that the loan was strictly for scientific purposes, Pratt immediately went to Holmes for advice and to form a partnership.[11]

Ravenel's lack of action during the summer indicates that he doubted claims of, or did not fully appreciate, the value of the phosphate rock. In handing the sample to Pratt, Ravenel likely sought a more expert opinion on the BPL level. Pratt would later write that Ravenel failed to "discover" South Carolina's phosphate rock because he used the wrong method of chemical analysis.[12] Given the developing rivalry that strained their relationship, Ravenel's avoidance of seeking Holmes' help was not surprising. But by giving the rock to Pratt, Ravenel lost an early and large advantage over the inevitable competition by failing to survey the territory, buy up the land, and start production before anyone else. With Ravenel's superior connections to Charleston capital, his failure to act on his alleged discovery suggests that he had some lingering skepticism of the rock's significance.

After his discovery, Pratt was cautious. Unless most of Charleston's phosphate rocks contained similar levels of BPL, Ravenel's rock would merely be a curiosity. Recalling Holmes' extraordinary knowledge of local geology and their wartime explorations, Pratt immediately visited his friend's office. When Pratt asked his old Nitre Bureau colleague if he knew the type of rock, Holmes replied, "Yes, as well as I know my own children" and volunteered comparable samples from his Ashley River plantation. Holmes warned him of Michael Tuomey's disappointing findings, only 16 percent BPL, but on August 10, 1867, Pratt obtained readings of over 55 percent BPL. The discovery was momentous for several reasons. First, Pratt had discovered the true chemical composition of much of the phosphate rock in the South Carolina lowcountry. Second, he had found the center of what would become the state's richest land-mining territory. Finally, he had established what would be the international standard (55 percent BPL) for the next quarter century.[13]

Pratt's findings put South Carolina's rock on par with the highest quality phosphatic material in the world. Correctly surmising that local land values soon would soar, Pratt told Holmes that they needed to determine the locations of the most abundant and accessible deposits as soon as possible. Pratt explored the phosphate rock outcroppings along the Ashley near Bee's Ferry with one of Holmes' assistants and found that "*it seems that there are no rocks in this section which are not phosphates!*" His enthusiasm was well founded. With an ample supply of rock near the surface, Pratt and Holmes began to see the discovery as the salvation of southern farmers and the lowcountry's prostrate aristocracy.[14]

On the same day Pratt explored along the Ashley, Holmes enjoyed his own moment of discovery and revelation. He received David Thomas

Ansted's essay "Lectures on Practical Geology," which described the geology of London's phosphate deposits. Ansted's work, for Holmes, "prove[d] the value of the [Ashley River] discovery" and "opened the eyes of scientific men here." Pratt, who already had the evidence he needed, felt that the book's significance was that it would reassure skeptical capitalists. Indeed, C. G. Memminger, the first potential investor Pratt and Holmes approached, said, "that book . . . is of the first importance in establishing the worth of your discovery, be careful of it." Holmes later credited Ansted's book for their eventual success in securing investors.[15]

Developing the Industry

With his BPL discovery, Pratt moved quickly to begin manufacturing fertilizer. Aware of Ravenel's ties to Wando, Pratt chose Holmes as his partner—a wise choice for several reasons. Besides being a scientist, planter, and expert on the Ashley River region, the well-known Holmes—and his close relatives—owned at least three thousand acres in the area. Although hounded by the federal government over his wartime activities, Trenholm remained a large landowner in the area. Even before they began soliciting for investments, Pratt and Holmes had "secured leases" on twelve thousand acres of phosphate lands along the Ashley. Charleston Mayor Peter Gaillard was the first landowner (one thousand acres) to commit to the new venture.[16]

Pratt and Holmes then spent six weeks in August and September 1867 canvassing Charleston in search of investment with which to start mining. They quietly circulated among gentlemen of means, hoping to attract investment without alerting others who might jump the claim by buying phosphate lands. Despite the all-but-tangible proof contained in Ansted's book, Pratt and Holmes found that Charlestonians were skeptical and hesitant to invest in their new enterprise. One local businessman remarked, "Dr. Pratt, do you, only recently come among us from Georgia, expect us to believe you, when you say that this material is worth and will bring $20 to $25 per ton, while men like Lyell, Agassiz, Tuomey, Ruffin, Holmes, Shepard, Hume and others, have known and handled it for twenty-five years? Excuse us, we cannot believe it." Generally stingy in nonagricultural

THANK YOU FOR JOINING US TODAY!

We continually strive to improve our programs. Your input will assist us.

BEAUFORT COUNTY LIBRARY
For Learning • For Leisure • For Life

This event: ☐ Exceeded my expectations ☐ Met my expectations ☐ Failed to meet my expectations

Did your knowledge of the topic increase as a result of attending? ☐ Yes ☐ No

How did you hear about this event? ☐ Library Newsletter ☐ Library advertisement ☐ Library email ☐ Newspaper ☐ Friend ☐ Other ______________

Do you have suggestions for future library events? ______________

Comments: ______________

____

Would you like to receive emails about Beaufort County Library news and events?

Email: ______________

Optional:

Your age	0-5	6-11	12-17	18-30	31-50	51-65	66-74	75+	**Gender**	Male	Female
Ethnicity	Caucasian	African-American	Latino	Asian	Other				**Your zipcode:**		

investment before the war and paralyzed by war and emancipation, low-country planter elites were even more cautious during Reconstruction.[17]

Charleston's aristocracy was in no mood to invest. Closely coupled to the city's economy, the Confederacy, and slavery, local elites lost businesses and real estate, and the city lost much of its infrastructure to Union shells and fires. Shortly after the war, Sidney Andrews described it as a "city of ruins, of desolation, of vacant homes, of widowed women, . . . of deserted warehouses, of weed-wild gardens, of miles of grass-grown streets."[18] Emancipation completed the disaster for Charleston's elites. The percentage of lowcountry wealth tied to slavery approached 80 percent in 1860, and African American freedom devastated land and property values, not to mention the investment in slaves. By 1865, the "old names" still held most of the wealth, but the mighty lowcountry planter class had taken a "staggering" hit, and their segment of the economy was shattered. After the war, Charleston's elites possessed smaller fortunes and were therefore more reluctant to invest in any venture, much less in a promising but unknown industry begun by scientists with little or no industrial experience.[19]

Pratt gained ample experience from local businessman John Commins in September 1867. Unsuccessful in his search for local investors and becoming "disheartened," Pratt turned to Commins, who claimed that his "business relations North" could raise $50,000 for the project. After demanding a pledge of confidentiality from Commins, Pratt disclosed his BPL analyses, the names of the land tracts, and the form of the leases. Pratt gave Commins a tour of the phosphate region, and Commins gathered rock samples from the Feteressa, Turnbull, and Goodrich plantations, which Pratt was then negotiating to lease. The next day, however, Commins abruptly ended their proposed alliance. Pratt soon learned that Commins had mailed the samples to a chemist in Philadelphia and had bought Turnbull and "tied up" Goodrich.[20]

Pratt and Holmes were unable to raise any genuine investment interest in Charleston, so they headed north. Contacting acquaintances from the North Carolina Geological Survey, Pratt obtained commitments from North Carolinians Washington Caruthers Kerr and E. Nye Hutchinson. Unfortunately, their investment was too small to start the enterprise.[21] Holmes sadly noted that "after six weeks of UNAVAILING

EXERTIONS" and having failed to raise enough southern capital, "we were compelled to leave home & resort to northern cities for aid" in mid-September 1867. Pratt and Holmes received funding for the trip from Charlestonian James T. Welsman, formerly of Trenholm Brothers of New York, a pardoned Confederate, and still intimately connected to Trenholm's empire.[22]

For southern patriots Pratt and Holmes, the trip to the North, a mere six months after the onset of Radical Reconstruction, must have been embarrassing and humbling. As did many white southerners at the time, the two men regarded the North and northerners with suspicion if not outright hostility. They would rather have enlisted Charleston men with Charleston capital to help resurrect this most southern of cities. The last thing Pratt and Holmes wanted to do was to deliver an industry with so much potential to northerners.[23] Holmes was especially stung when Charlestonians questioned his sense of patriotism in exploiting the discovery. Edward Willis, in 1872, was not the first local businessman to claim that Pratt and Holmes stole northward and "disclosed" their "secret" and "plans" to northerners. For Willis and others, Ravenel's Wando Company, started and maintained with local capital, provided the obvious contrast to Pratt and Holmes' alleged disloyalty. Of course, with Charleston's economy largely based on trade with outsiders, this resentment represented the height of hypocrisy.[24] Nonetheless, the accusations haunted Holmes for the rest of his life.

Other Charlestonians were sympathetic to Holmes and Pratt and acknowledged that northerners controlled most of the nation's entrepreneurial capital. While still proud of their region and eager to rebuild the economy quickly, these men justified Holmes and Pratt's journey north by arguing that the postwar period was the worst possible time to tap the South's capital resources. A commiserating author conceded that southern whites had "absolutely no disposable capital" and that those with any were "naturally chary of investing . . . in an enterprise so novel, and at the first sight, so chimerical."[25] There seemed to be little recourse to going north to Philadelphia for help.

For Holmes, the fact that Philadelphia was familiar territory mitigated the pain of going begging to the "enemy." Indeed, although Washington and Baltimore were geographically closer to South Carolina, Philadelphia was the northern city with which nineteenth-century Charlestonians shared the deepest bonds of family, friendship, education, and business. During the Revolutionary War, Charleston's exiles fled to Philadelphia. In the antebellum period, young southern gentlemen traveled to Philadelphia to enjoy superior medical training and the city's pro-southern atmosphere. Despite the ascension of Republican politics, postbellum Philadelphia remained home to an elite that shared with southern planters a "socially reactionary" outlook (especially toward abolitionists) and membership in "the American aristocracy." The relationships between elites in the two cities survived the Civil War and played a significant role in the development of South Carolina's phosphates.[26]

Holmes' connections with the city and its scientific community dated back to the 1840s and 1850s, when, as a corresponding member of the Academy of Natural Sciences of Philadelphia, he knew scientists Samuel G. Morton, Joseph Leidy, and Samuel H. Dickson. Holmes' scientific credentials and ties aided his search for capital and gave his proposed venture instant credibility. Holmes and Pratt soon struck a deal with investors George T. Lewis and Frederick Klett. An industry insider with a European PhD, Klett headed the "innovative" sulfuric acid and superphosphate manufacturer Potts & Klett, one of Philadelphia's top four fertilizer companies.[27] Lewis was a prominent chemical manufacturer in Philadelphia and, according to Pratt, "among all American Capitalists was the first to appreciate the value of our phosphates." The four men organized the Charleston, South Carolina, Mining and Manufacturing Company (CMMC) in September 1867.[28] Together, Lewis and Klett helped Holmes and Pratt create a lasting business by contributing abundant capital, sound business structure, and existing markets.

CMMC

Combining expertise and local knowledge with ample funding, the new mining company held substantial advantages over later entrants into the industry. Pratt and Holmes knew the location of the best-known rock deposits, and their northern investors provided the capital to lease or buy much of the available land along the Ashley River. CMMC pursued land in a market depressed by war, death, emancipation, taxes, poor harvests, and a declining rice industry. Many plantation owners were willing to lease or sell family lands. Although a Baltimore newspaper labeled CMMC's early leasing and buying spree an unsuccessful attempt at a phosphate "monopoly," the company's outlays for land, along with similarly generous expenditures for machinery, labor, and transportation, enabled CMMC to dominate, if not monopolize, land mining.[29]

Representing other investors, Lewis joined Philadelphia merchants Samuel Grant Jr. and Samuel F. Fisher in a fall 1867 trip to the Ashley phosphate territory. The group stayed at Drayton Hall plantation and conferred with Pratt and Holmes. Aside from being "the only decent habitation in that part of the country," Drayton Hall was just a few miles northwest of Bee's Ferry and Holmes' former plantation, Springfield, and an excellent base of operations from which to assess the deposits. According to a visitor then staying at Trenholm's nearby Vaucluse plantation, "it did not take these gentlemen long to look over the field and buy up thousands of acres."[30]

Securing phosphate-rich lands was the top priority for Holmes, Pratt, Klett, and Lewis. Headed by the two southerners, CMMC's negotiations with landowners formally began in October 1867 and concerned properties in St. Andrews Parish near Bee's Ferry. By October 20, Pratt and Holmes had leased six properties, most of which had access to the Ashley. With no railroad spurs near the first mining areas, the river was crucial for transporting the rock to Charleston. CMMC's first plantations totaled 4,455 acres, and Pratt and Holmes initially controlled them with ten-year leases. Klett and Lewis bought a one-third share of the leases with payments of $33,333 to Pratt and Holmes over the next four months. Pratt and Holmes would continue to own the remaining two-thirds shares. In addition to cash payments, CMMC, according to Holmes, paid 10 percent of the value of rock mined to the landowners in 1870.[31]

CMMC's earliest surviving documents reveal the nature of the land transactions. Memorandums dated mid-October 1867 were legal agreements between the landowners and Pratt and Holmes and did not include any mention of the company. Klett and Lewis' capital flowed through the scientists to pay for the leases. Aside from their geologic, geographic, and chemical knowledge, Pratt and especially Holmes added a personal touch to the land-acquisition process. Local landowners negotiated with the familiar scientists, not a faceless company dominated by northern investors. Initially, landowners preferred leasing over selling, and some leases may have included, as did later such documents, provisions prohibiting mining near the plantation houses and gardens. Finally, and foreshadowing CMMC's efforts to secure the entire region, the documents also included provisions for controlling "other similar lands adjacent."[32]

By mid-December 1867, CMMC moved beyond the leasing stage and began to buy land. On December 13, Pratt and Holmes, "for the benefit of" the company, bought Hickory Hill plantation and, five days later, The Oaks. By February 13, 1868, Pratt and Holmes had already acquired, or were negotiating to buy, eleven plantations. Of the 10,749 total acres purchased with $103,180, at least three of the properties—Ashley Ferry, The Oaks, and Maryville—initially had been part of the October 1867 leasing agreement. The three property owners—Holmes, M. G. Ramsey, and David W. Lamb—may have witnessed the physical destruction of strip mining on their plantations and, realizing that the property might never regain agricultural or aesthetic value, decided to sell. Conversely, CMMC may have sensed more profit potential or potential competition and bid higher for the lands.[33]

An obvious feature of the February 1868 purchases was that the price per acre varied dramatically, more so than Pratt and Holmes' preliminary investigations of the ten thousand acres would warrant. Not surprisingly, the insider Holmes got the best price, and 50 percent in cash, for his land. "Colonel" Joseph A. Yates got the worst price, but within the year he became superintendent of the company, possibly through an unwritten part of the deal. Another factor in land price was the presence of extensive marshlands bordering the Ashley. In order to secure access to the river, CMMC had to buy marshlands that, in most cases, were difficult to mine and thus cheaper.[34]

CMMC's 1868 land transactions were a mixed process of personal and corporate exchange that revealed the declining power of Pratt and Holmes within the company. Of the eleven proposed or executed bonds securing the credit purchases, six directly involved the company, while the scientists pledged the remainder. The two men were to immediately "convey with proper renunciations of Dower" the lands under their names to the company and became indemnified "against all liability thereof." In contrast to

Pratt and Holmes' limited liability, the other seven CMMC investors assumed unlimited liability for the bonds. That Pratt and Holmes enjoyed limited liability in the partnership confirmed their lack of investment in the company. Their primary contributions to the success of CMMC—the original idea and local knowledge—were weak footholds in the company, which would quickly erode after the land purchasing stage.[35]

By early 1868 then, CMMC had bought or leased most of the best phosphate lands in the area. Within a year of its founding, CMMC owned about ten thousand acres on both sides of the Ashley and controlled through long-term leases another ten thousand acres. The buying continued for decades and into other counties and states. The acreage easily dwarfed the acquisitions of all other land-mining companies combined and fundamentally altered the region's future development.[36] The early and ample land purchases helped to establish CMMC's position as the dominant company in the state.

Beyond buying land, CMMC expanded its investor base in February 1868 with other southerners joining the Charlestonians within the Philadelphia-dominated company. Unable to shoulder the entire investment for Pratt's idea six months earlier, North Carolinians Hutchinson and Kerr finally bought into CMMC and brought fellow residents Thomas J. Summer and Robert Frederick Hoke with them. Like Holmes and Pratt, the men were conservative southerners eager to promote industry and agriculture in the South. Whether CMMC brought on board the North Carolinians—and especially Hoke, a former Confederate major general who served with Robert E. Lee—in a public relations move to stifle lowcountry objections remains unknown. Regardless of CMMC's motivation, muting southern fears of invading carpetbag capital served the company well during the volatile political atmosphere of 1868. No popular animosity toward the company developed.[37]

CMMC tallied another significant acquisition when Trenholm joined the company's board in February 1868. The revered Charlestonian was quietly active in the land-mining and fertilizer-manufacturing industries, but the extent of his involvement remains unclear, likely due to his legal troubles. Ever the booster of Charleston, the South, and his fortunes, Trenholm joined the CMMC's board for symbolic and financial reasons. His involvement fully sanctioned the new company and industry. Rumored to be hiding a fortune in blockade-running profits from federal prosecutors but living modestly after the war, Trenholm was not merely a symbol. For the businessman and landowner, the profit potential for the new industry was hard to ignore. Trenholm's antebellum ownership of Ashley River tracts, including the Vaucluse, Springfield, and Ashley Hall plantations, suggests that he retained substantial interest in phosphate-rich lands. Although the CMMC board was his last visible, direct footprint in the land-mining industry, Trenholm's invisible hand could be inferred in the active participation of family and business associates and in his related businesses. G. A. Trenholm and Son became the largest firm in Charleston doing general commission business chiefly in phosphate and fertilizer. Besides his brother-in-law Holmes, Trenholm's sons William and Frank, as well as business associates James T. Welsman, Andrew G. Magrath, and Edward Willis, were active in the industry over the next decade.[38]

CMMC's purchases and Republican tax policies set off a land boom in late 1867 and early 1868 along the Ashley and especially among the struggling planters in St. Andrews Parish. Many of the plantations below Middleton Place had little agricultural value before or after the war and merely served as traditional family seats. Republicans made an aggressive tax policy the core of their program in Reconstruction South Carolina and specifically targeted unused real estate and property. In the lowcountry, the policy served as a catalyst for the land-mining industry. Large landowners now had strong economic incentives to make all of their land productive or risk losing it to sheriff's sales. For many of those owning phosphate-rich lands, the only options were to mine, lease, or sell the land. According to Holmes, local plantations that sold for two dollars per acre before the phosphate discovery "immediately advanced to twenty" afterward. Several owners, including the Draytons of Drayton Hall, initially leased and then eventually mined the land themselves. C. C. Pinckney Jr. later mined his Runnymede plantation extensively. And Williams Middleton began a second career.[39]

CMMC's land purchases had an immediate impact on Middleton Place. Having lost more than one hundred slaves to emancipation and the great mansion to General William T. Sherman's torch, Middleton had rented out the estate before moving back to the ruins in 1867. Borrowing money from his sister Eliza M. Fisher in Philadelphia, Middleton attempted to grow rice but found it difficult without slaves. As early as January 1868, Middleton perceived that he could raise another crop at Middleton Place. CMMC's leases and purchases led to a lively round of letters and telegrams between Williams Middleton, Eliza's husband, J. F. Fisher, and Williams' nephew John I. Middleton Jr. in Baltimore concerning land mining.[40]

Closely following the latest rumors and news about CMMC in Philadelphia, J. F. Fisher bombarded Middleton with advice in January 1868 on the value of phosphate-rich Middleton Place. Fisher warned his brother-in-law not to sell any land, sign any deeds, nor trust anyone, because, he speculated, demand was outstripping supply and, therefore, the property must be worth millions. He heard that "phosphate speculators" bought nine square miles on the Ashley valued at more than "ten millions." Another missive reported that speculators bought fifteen thousand acres and sought to establish a monopoly. Fisher then sent Middleton five hundred dollars to enable him to hold on while he negotiated the phosphate boom. Fisher also counseled Middleton to evade the execution of an existing contract with local businessman George S. Cameron, who was backed by Pennsylvania's U.S. Senator Simon Cameron. Fisher was working with a Mr. Coxe to free Middleton from "Cameron's clutches."[41] Meanwhile, John I. Middleton Jr. introduced Williams to Robert Turner and Charles J. Baker, Baltimore fertilizer dealers who were interested in a mining venture on Middleton Place. Turner and Baker offered to provide "*all the funds required*" if Middleton supplied the land and supervision. The three soon agreed to terms and formed the Ashley Mining and Phosphate Company. By mid-April 1868, Middleton Place had yielded one hundred tons of phosphate rock, and Williams Middleton still controlled the plantation.[42]

The CMMC-induced land boom soon spread beyond St. Andrew's Parish. In March 1868, John Commins advertised in the *Charleston Daily Courier* for phosphate samples from local landowners. By April, the newspaper's editor declared that the "phosphate fever" was producing a land stampede similar to an oil rush and that the new fertilizers would "work miracles" for the state. Community attitudes had evolved in a short period of time from "the extreme of skepticism" to "the extreme of exaggeration and credulity."[43] The money flowing from Philadelphia into Charleston ignited hope, greed, and entrepreneurship in a lowcountry aristocracy that was beginning to envision an economy and society after slavery.

Figure 2. CMMC rock washer at Lamb's, 1877. By permission of South Caroliniana Library, University of South Carolina, Columbia.

Figure 3. CMMC workers and manager at Lamb's, 1877. By permission of South Caroliniana Library, University of South Carolina, Columbia.

After its initial land-buying spree, CMMC continued to build dominance within the industry by tapping its ample capital reserves for the plant and equipment. Throughout 1868, the company bought machinery and directed the construction of buildings, wharves, and tramways on the newly acquired lands in preparation for large-scale mining. The flow of CMMC's funds did not slow, despite the fact that the company, originally chartered in December 1867 with one hundred thousand dollars, spent more than that on land leases and purchases in the next two months. CMMC belatedly amended its charter in March 1869 to six hundred thousand dollars capital with the "privilege" of increasing its capital to one million dollars, which the company did (again belatedly) in May 1869. The company's reluctance to state its true value likely reflected the owners' desire to avoid the cumbersome chartering process and reluctance to flaunt the company's "foreign" wealth.[44]

Having secured land and constructed the plant, CMMC built an efficient transportation and processing operation. Acquired in January 1868, David W. Lamb's property provided an ideal location for organizing mining operations and processing the rock. "Lamb's" was on the Ashley, accessible to railroad lines, and centrally located amid CMMC's new properties. From the mining pits, laborers moved the crude rock on trams to riverside wharves and then to Lamb's via lighters or other shallow-drawing vessels. After workers and machinery crushed and washed the rock, CMMC's boatmen shipped it down the shallow Ashley to the deepwater Cooper River wharves where longshoremen loaded CMMC rock onto oceangoing vessels bound for northern ports. Lamb's featured first-rate processing machinery, including two northern-built washers run by a forty-horsepower engine with one-hundred-ton daily capacity. The extensive processing works and wharves at Lamb's were operational in April 1868, and by 1870, the company had spent $150,000 on buildings, mills, wharves, machines, locomotives, and rails at the location. CMMC's spending on Lamb's and nearby facilities augmented the company's early territorial advantages and extended CMMC's industry dominance. While several other local mining companies shipped unwashed rock to northern ports and saw the filthy cargoes thrown away, CMMC sent well-processed rock and began to build brand-name recognition. Since CMMC shipped the majority of the state's rock, the company helped build farmer loyalty for South Carolina rock as well.[45]

While buying land and building the plant, CMMC marketed its product. The new company sent sixteen barrels of phosphate rock on December 19, 1867, to George T. Lewis in Philadelphia. Lewis sent a portion of the rock to Coates & Company of London and the balance to Potts and Klett. Generating a "great excitement in the fertilizer world," Coates & Company forwarded portions of the rock to "distinguished chemists" in Switzerland, Sweden, Denmark, Austria, Prussia, France, and England. In Philadelphia, Potts and Klett incorporated their sample into the first superphosphate fertilizer manufactured with South Carolina rock. CMMC may have had other buyers for the rock in Philadelphia, but through director Klett, the new company had an established market for its new product. In an international market where suspicious farmers developed brand loyalties to the finished product as well as its raw materials, CMMC's entrance into the trade was relatively quick and painless—another enormous advantage.[46]

Having built the dominant land-mining concern in the state, CMMC's directors belatedly established the company's first international headquarters in the back of Holmes' Book House at 60 Wentworth Street in Charleston. Such unhurried movement from those who dramatically purchased land and machinery demonstrated the priorities in this extractive industry. Investment, land, and machinery were crucial elements, but a home office and bureaucracy were not.[47] The location of CMMC's headquarters also reflected the initial place of Charlestonians in the new company. The Philadelphia backers elected Holmes as CMMC's first president, Pratt as chemist and superintendent, Holmes' son-in-law, Arthur H. Locke, as secretary, and the firm Pressley, Lord & Inglesby as its lawyers. Sensitive, with good reason, to being perceived locally as carpetbagger capital, CMMC's Philadelphians shrewdly anointed Charlestonians as the company's most visible representatives during the company's first and largest stage of land acquisition. Pratt and Holmes may have suggested such a course, given the simmering controversy over their 1867 trip to Philadelphia. Dependent on local labor and businesses and creating a new industry from scratch, CMMC could ill afford local animosity against a group of northerners buying Charleston's plantation lands shortly after the war.[48]

Despite an unimposing office and staff, CMMC was active in pioneering public relations through the local press. In eleven *Courier* articles during 1867–68 and one in the *Charleston Mercury*, Holmes, Pratt, and CMMC cultivated boosterish press coverage. By contrast, CMMC's only real competitor at the time, the locally owned Wando Company, did not aggressively woo reporters and consequently reaped only three brief articles during the same span. Wando's leaders likely discounted the need to create a favorable public image because they felt the company already had one. Still defensive about their trip to the North, however, Pratt and Holmes actively solicited newspaper coverage. Twice in October 1867, the two scientists surveyed potential mining areas with *Courier* reporters. In November, the *Courier* announced CMMC's birth in glowing terms, emphasizing Charlestonians' roles with no mention of Philadelphia. Pratt and Holmes placed an ad in the *Courier* in December offering to give "advice and instruction to all persons interested in mining, improved agriculture, chemical analysis or manufactures." In April 1868, the *Courier* ignored Wando's pioneering shipment and instead trumpeted CMMC's as "the first" commercial shipment. Holmes, Locke, and Yates lavished attention on the *Courier* and *Mercury* editors during a visit to CMMC's operations in July 1868. The editors responded with two lengthy articles minimizing the company's northern ownership. Subsequent reporting in 1868 revealed CMMC's eagerness to share with reporters the most minute details of the company's operations.[49]

CMMC's policies of journalistic openness and southern visibility paid off handsomely. South Carolinians did not appear to view CMMC as a northern company forming a colonial relationship in the "prostrate" South. Rather than emphasizing the Yankee capital, control, and profits within CMMC, local reporters devoted most of their coverage to the exciting new *southern* company and industry. The visibility of Holmes, Pratt, and Locke during the land-acquisition phase proved to be an adept move. Southerners read that fellow southerners discovered the substance on southern soil and that they ran the company. Southerners, not northerners, appeared to be buying the land. Southerners supervised freedmen miners on former plantation lands. The well-publicized participation of the two southern scientists gave the Yankee-financed company a southern flavor and did much, at least in the lowcountry, to assuage southern pride.

GREATLY INCREASED CROPS

BY THE USE OF

The Best and Most Reliable Home-Made Fertilizer,

"THE WANDO."

MANUFACTURED BY THE

WANDO MINING & MANUFACTURING COMPANY,

At their Works in Charleston, S. C.

WM. C. DUKES & CO.,

GENERAL AGENTS,

No. 1 South Atlantic Wharf.

Figure 4. Wando advertisement, 1870. By permission of South Caroliniana Library, University of South Carolina, Columbia.

But the Charlestonians' prominence in CMMC masked the real power of the northern investors. Most of the directors lived in Philadelphia, and the company established a branch office and agency there. Philadelphia merchant Jesse E. Smith was the largest shareholder. After the land purchases, President Holmes became more of a figurehead than a driving force at CMMC. Similarly, Chemist and Superintendent Pratt languished in a company that produced no commercial fertilizer. Less than two years after launching the company, Holmes and Pratt dropped out of active participation. By May 1869, the company's headquarters had moved to Philadelphia, and Smith had become president. Locke manned the small Charleston office, and Superintendent Yates organized the labor and transported the "crop," two jobs better suited to a planter than a scientist. While remaining on the board of directors, Pratt and Holmes moved on to other pursuits—fertilizer manufacture and retirement, respectively.[50]

Wando

The Wando Fertilizer Company reflected the characteristics and culture of a truly southern-based company. While Pratt and Holmes helped CMMC become a land-mining behemoth and then left the northern-dominated company, Ravenel and partners continued to develop Wando with southern investment and talent. "Local capital" meant less capital, which the company initially listed at one hundred thousand dollars. As opposed to CMMC's emphasis on land and mining, Wando's directors consistently focused on the fertilizer business, with mining a side venture to partially supply the factory. Certainly, the directors could have used more capital for the fertilizer business, but their plans did not include a mining empire. In contrast to its capital, Wando's talent was more than adequate. With Ravenel as chemist, Ebaugh as superintendent, John R. Dukes as president, and William C. Dukes and Company as general agents, Wando soon created an effective research department, management team, and sales force.[51]

Throughout the rest of the century, most of South Carolina's land-mining and fertilizer-manufacturing companies followed Wando's organizational formula. Companies often included the president, chemist, superintendent, and agents on the board of directors, and family connections between the president and the agents, or among board members, were common. Familial links were rare, however, between the often elite board members and the superintendents. Chemists frequently worked as consultants, and superintendents and their assistants had the only full-time, managerial positions. In most cases, a superintendent worked on-site and reported to the president, who was usually a factor, banker, or lawyer. The company often borrowed a small clerical staff from its law firm.

Agents had greater roles in fertilizer manufacturing than in land mining. Most agents were cotton factors, which proved helpful in fertilizer marketing. Closely allied with Wando's president, the factorage firm of William C. Dukes & Company was Wando's first general agent. Dukes marketed, sold, and shipped Wando's fertilizer to local farmers and agents in neighboring states. Although mining company agents, such as CMMC's B. S. Rhett & Son, may have sold some ground rock to local farmers, their primary responsibility was to ship the mined product to Philadelphia and other manufacturing centers in the Northeast. Not surprisingly, land-mining companies often appropriated the agent position.[52]

Wando's small, family-oriented business structure resembled nineteenth-century French business culture, in which owners emphasized security over risk and inhibited the use of outside capital and new machinery or methods. In contrast to the more profit-focused British, the French viewed firm and family together to be an end in itself. Business activity in Continental Europe, especially France, tended to be a "class activity," in which "aristocratic industrialists" maintained their strongest ties to agriculture and dabbled in industries like mining that depended on landownership. Aspects of the French pattern resonated in the land-mining and fertilizer-manufacturing industries.[53] The French connection is especially significant considering the influence of Huguenots in Charleston's history. Although it is difficult to gauge precisely their impact on the city's business culture, we must acknowledge their central place in the lowcountry. Many Huguenot descendants—most notably the Ravenel, de Saussure, Guerard, Huger, Peyre, Porcher, Prioleau, and St. Amand families—played pivotal roles in land-mining and fertilizer companies.[54] The Huguenot influence, along with a business model derived from cotton factoring, made the locally owned mining and manufacturing businesses unique in style and culture.

Wando's progress immediately after the phosphate "discovery" fit within Charleston's French-influenced business culture. Its scientific expert,

Huguenot descendant Ravenel, was as cautious as northerner-dominated CMMC was aggressive in exploiting the phosphate windfall. Wando bought 859 tons of "phosphate rock" from Navassa Island in the Caribbean as late as November 1867. Previous historians have explained Wando's seemingly excessive caution by speculating that Pratt hid "the secret" from Ravenel and then smuggled the information to Philadelphia. Available evidence refutes this conspiracy theory. The alleged secret was "generally known" by early fall in Charleston. October press coverage of fossil discoveries would have confirmed any suspicions of a land rush.[55] The more likely reason for Wando's delay in using local rock was that CMMC outbid it for mining lands. Alarmed at CMMC's bold initiatives, Wando's directors probably ordered Navassa rock as a supply bridge until they could buy mining territory. Wando's factory had begun production in November 1867, and the company could not risk running out of rock. But Wando's importation of phosphate and sulfuric acid did not make long-term business sense. In December, Wando acquired phosphate beds on the Ashley near Bee's Ferry and began manufacturing fertilizer "almost entirely of home material." Although it would never rival CMMC's rock production, Wando made history when it vertically integrated its manufacturing with mining. The *Courier* praised the "home institution," noting that the cheaper fertilizer would "keep many thousands in money here instead of continuing the old plan of buying abroad"—meaning the North, seemingly a foreign country at the time.[56]

Despite Wando's primary focus on fertilizer, it was the first company to export South Carolina phosphate rock. Wando sent small shipments in December 1867 to New York fertilizer merchant George E. White and to a Dr. Clements of Baltimore. Lacking CMMC's insider contacts in Philadelphia, Wando sought to build market share in two other fertilizer centers. Wando also upstaged CMMC the following spring. In April 1868, Wando produced the first batch of superphosphate made in Charleston containing local rocks and shipped the state's first commercial cargo of phosphate rock (one hundred tons) to Baltimore. CMMC followed a few days later with a three-hundred-ton shipment to Philadelphia.[57]

Unable to match CMMC's capital, Wando nonetheless emerged as a successful miner, thereby demonstrating that the former's aggressive land and plant purchasing did not raise insurmountable entrance barriers.

CMMC would dominate the land-mining industry throughout the rest of the century, but the part played by smaller companies such as Wando would not be insignificant. Large and small firms shared in the industry's development, specifically with innovation in technology and organization, creation of brand-name recognition, and building of local infrastructure and foreign contacts and markets. Both companies' first shipments made an immediate impact on the northern fertilizer industry, especially after northern chemists gave their blessings to the "true Bone Phosphate-rock." Holmes wrote that CMMC's first cargo "caused no little excitement in mercantile circles," and that, before long, "Phosphate fever became EPIDEMIC" in Philadelphia, New York, and Baltimore.[58]

Holmes' enthusiasm masked some early difficulties with South Carolina's phosphates, especially processing and purity. Without proper washing and drying, land-rock shipments contained moisture and impurities that increased the weight of each shipment but decreased the BPL percentage. Both problems drove up costs for northern and European buyers, leading to reluctance to buy the state's rock. In the brand-conscious and fraud-weary fertilizer business, these deficiencies threatened the new industry's reputation. Many of Charleston's mining companies, including Wando, chose short-term profits over marketing and ignored the issue for a decade. Early miners merely sun dried the rock, which failed to remove most moisture. Washing consisted of "a rough scrubbing with hand brushes in a convenient creek," a method sufficient for rock found in sand but inadequate for the majority of rock found in dirt and clay. Buyers had to have shipments "mined out of" cargo holds and rewashed—unacceptable hassles, especially to European buyers, who expected uniformly sized and washed rock of at least 55 percent BPL. Indeed, sloppy preparation led Europeans to form a "strong prejudice" against South Carolina land rock, which the industry was never fully able to overcome.[59]

Better financed than Wando, CMMC confronted the washing problem by building Washer No. 1 at Lamb's. Built in 1868 on the banks of the Ashley, the elevated structure had a thirty-five-horsepower engine and a double-screw washing machine, similar to devices used in Ohio to process iron ore. CMMC workers dumped the rock onto a "trough," where No. 1's "turning cylinders" propelled and washed it in river water. The rocks then tumbled into a "great tub," where, an admiring reporter claimed, they underwent a "thorough cleansing." A big step forward, CMMC's Washer No. 1 had a capacity of two hundred tons per day but still did not meet European expectations for cleaning. A year later, CMMC improved the previous design with Washer No. 2, adding more elevation for more efficient dumping and washing. No. 2 used gravity to keep the rock and water moving faster to increase capacity. The wooden two-story washer surrounded by ramps and rails became a standard feature of the land-mining industry.[60] Sketches of Lamb's over a decade later reveal a similar process. Part of a structure approximately thirty feet tall, a steam engine propelled gears and conveyor belts, and wooden viaducts carried water and rock throughout the structure. Supervised by a white man, about fifteen black men with shovels kept the rock moving. Men and machines jostled, broke, and doused the rock and then loaded it back onto wheelbarrows.[61]

Just down river from Lamb's, Wando built a two-story washer at its mine as early as October 1869. A smokestack indicates that Wando's workers may have partially dried rock near the mine, but they moved the still-wet rock down the Ashley and up the Cooper to the company's fertilizer factory, where others dried it in "two large furnaces and ovens." Vertically integrated with mine and factory and a local producer, Wando could afford to dry at the place of manufacture. As the industry matured, most mining companies dried the rocks near the mine, immediately after washing and before transportation.[62] Wando used most of its locally mined rock in its fertilizer, which began to make its mark in the Southeast by June 1868. Farmers growing corn and cotton in Christ Church Parish reported success using Wando's product, while observers described non-fertilized crops as "backward and dwindling." Later that year, a Wilmington, North Carolina, newspaper noted that Wando's fertilizer had become a "favorite" with farmers.[63]

While successful in mining and manufacturing, Wando's directors could not ignore CMMC's aggressive expansion. In September 1868, the Dukes reorganized Wando Fertilizer Company as Wando Mining and Manufacturing Company, a better-funded version of its predecessor. The new Wando added more backers and directors and had the "privilege" of increasing its capitalization to five hundred thousand dollars. A testament to the change in perception about the new industries, finding local investors had become easier in the year since Holmes and Pratt's fruitless search.

For South Carolinians able to commit capital, land mining and fertilizer manufacture now appeared to be economically profitable, agriculturally complementary, and socially acceptable.[64]

Wando's new directors had ambitious plans, including a new factory and expanded mining. They also hoped to continue manufacturing with local phosphates, increase production, and expand sales territories beyond the Carolinas and Georgia.[65] Despite mining success, most of Wando's resources went into fertilizer production. During 1867 and 1868, the company's board worried about a CMMC monopoly that might restrict Wando's access to local rock, so they bought some of the acres the mining behemoth had not yet secured. Wando's mine production jumped from a small sample in 1867 to 2,279 tons in 1868. But by 1869, the board perceived the threat of monopoly easing as other firms entered the field. Wando's totals dropped sharply to 241 tons in 1869 and 562 tons in 1870. Reassured that CMMC had some competition and that local rock was available, Wando stopped mining by 1871, bought rock from the many local producers, and focused on its core business, fertilizer.[66]

Other Entrepreneurs

Favorable press coverage and the promise of ample phosphate markets spurred a variety of local and northern entrepreneurs to enter the land-mining business by 1870. Among those joining CMMC and Wando in the phosphate fields were locally owned companies Ashley Mining and Phosphate (Middleton Place), Atlantic Phosphate (Livingston Farm), Chicora Mining and Manufacturing (Filbean Creek), Palmetto Mining and Manufacturing (Spring Farm), and F. H. Trenholm (Drayton Hall). Major George T. Jackson (Hard Farm) and A. J. and O. A. Moses (Massot Farm) mined north of the Charleston Neck and between the Ashley and Cooper Rivers. Another local firm, Farmers' Fertilizer Company, had "control" of lands at "Phosphateville" and Shipyard Creek along the Cooper. McDovel and Drane mined for George S. Cameron along the Stono River near Charleston. Still others found phosphate rock south of Charleston County. In Beaufort County, Williman's Island Phosphate mined near Bull River and North Wimbee Creek, while Oak Point Mines worked Kean's Neck between North and South Wimbee Creeks. Along the Edisto River in Colleton County, W. L. Dawson mined near Parkers Ferry, and Frederick Fraser's Horseshoe Mining operated near Jacksonboro.[67] Most of these would prove to be short-lived ventures, but the numbers indicated an explosion of local interest in phosphate land mining.

Sensing the importance of the South Carolina discoveries, northeastern fertilizer producers joined local miners in the lowcountry's phosphate fields. George S. Cameron came from Philadelphia to mine near Middleton Place. William L. Bradley of the Boston fertilizer dynasty mined at Eight-Mile Pump for the locally owned Carolina Fertilizer Company, thus beginning his family's forty-year involvement with the state's three phosphate-related industries. Fertilizer merchant J. B. Sardy, with offices in New York and Savannah, operated Ashepoo Mines in Colleton County to supply his Charleston fertilizer works. Boston-based Pacific Guano Company mined Chisolm's Island near Beaufort to supply its fertilizer factories.[68]

Receiving financial support from local, regional, and national sources, the new companies represented a blossoming industry led by a diverse group of entrepreneurs. A sixty-one-year-old former slave owner, Williams Middleton supervised Ashley Mining's operations and received financial backing from Baltimorians Baker and Turner. Atlantic's board included some of the most prestigious lowcountry names—Porcher, Pinckney, and de Saussure. Twenty-five-year-old A. D. Estill served as president and superintendent of Chicora, which received support from local businessmen. Farmers' directors hailed from Abbeville, Charleston, Chester, Columbia, Fairfield, Greenville, Sumter, Union, and Charlotte (N.C.) and elected Charlestonian William G. Whilden president. Jackson reemerged in Charleston's fertilizer industry for the first time since dissolving his partnership with Charles U. Shepard Sr. in 1861. Besides operating Palmetto, Thomas Eason owned Eason Iron Works and was an inventor; he built Ashley's washer. Like Wando, several of these companies eventually abandoned their mines, bought from local producers, and focused on fertilizer manufacture.[69]

The biggest difference between the northern-based companies and their local counterparts was capital. In 1870, the Pacific Guano ($1 million), CMMC ($1 million), and Bradley ($500,000) companies dwarfed their locally based competitors. Wando ($300,000), Atlantic ($200,000), and Farmers' ($200,000) led the southern division, with Chicora ($70,000) and the others trailing far behind. In addition, while northern firms listed collected capital, southern companies often recorded only pledged capital. Chronically short of funds, several of the local mining entrepreneurs spent most of their money on fertilizer manufacturing and shared or sublet expensive mining equipment, such as washers. For example, Wando and Atlantic devoted only a small percentage of their capital to mining, while CMMC directed almost all to its mines. Although Pacific Guano was foremost a fertilizer manufacturer, its Chisolm's Island operation represented a fundamental shift in the company's supply chain for raw material and therefore received abundant funding from the international fertilizer giant.[70]

The best basis for comparison was phosphate land-mining production. During the industry's early years, CMMC dominated its nearest competitors. In 1868, CMMC shipped almost 50 percent more rock than F. H. Trenholm and almost twice as much as Wando, the two local leaders. Mining companies sent most of the rock to domestic ports, and Pacific Guano and Wando consumed much of their production. Only two companies sold rock abroad in 1868, shipping a mere 1.7 percent of the state's total production to foreign ports. In 1869, CMMC doubled its tonnage, producing more than its next four rivals combined. At least six companies sold rock abroad in 1869, increasing exports to 13 percent of the state's rock total. In 1870, CMMC increased production by 50 percent and remained comfortably ahead of all other producers. Having at least doubled production in each of the first three years, the South Carolina land-mining industry had become a viable supplier of raw materials for the international fertilizer industry.[71]

F. H. Trenholm's experience as a mining entrepreneur illustrates the difficulties faced by small local producers, even those with prominent backing. Frank Trenholm leased part of the Drayton Hall plantation from Dr. John Drayton in 1868. With his father, George A. Trenholm, and uncle, Francis Holmes, on CMMC's board, Frank's mine was a training ground for the young man and another component of the Trenholm empire. In his first year, he sold 1,680 tons to "S. Grant Sr.," likely associated with Samuel Grant Jr., a CMMC investor, and two hundred tons to the Sulphuric Acid and Super-phosphate Company, a local fertilizer producer linked to CMMC. By 1869, Trenholm's mine ranked behind only CMMC in production.[72]

Frank Trenholm's problem was that while CMMC bought its land or secured ten-year leases, Drayton only offered him a two-year lease at eight hundred dollars per year plus dividends based on rock sales. For mining entrepreneurs, a long lease was necessary to recoup initial investments on processing and transportation facilities. Trenholm had to build "the necessary machinery" to expedite mining the entire deposit, but he faced losing all the fixed plant and improvements after two years. In November 1868, he wrote Drayton about his plans to spend forty-five hundred dollars enlarging the mining plant with two washers, an engine, and a pump, and another two thousand dollars for a railroad extension and more tram cars, wheelbarrows, picks, and shovels. Hoping this impressed his landlord, Trenholm asked Drayton for a guaranteed two-year extension or, preferably, a ten-year lease. Anxious about the competition, he wrote Drayton, "I am not in a position to make any offers to supply for want of proper machinery which I don't dare erect on my short lease." Without a better lease, Trenholm threatened to "give up 'Drayton Hall'" and "try & get land elsewhere."[73]

Trenholm's bluff became Drayton's opportunity. The absentee landowner had already paid one-third the cost of Trenholm's machinery, and his lawyers, Loftus C. Clifford and J. Fraser Mathewes, advised him to force Trenholm out. As land values rose throughout 1868, the lawyers put off Trenholm's appeals while they shopped for a better deal. In December, Clifford and Mathewes signed CMMC to a five-year lease starting after Trenholm left. The new lessee would pay Drayton one dollar per ton and a cash advance of $750, dig a minimum of two thousand tons per year, build a wharf and buildings to house machinery, and avoid disturbing the "Yard and garden grounds." Upon the expiration of the lease, Drayton would own all except the machinery.[74]

Mathewes informed Trenholm on January 11, 1869, that his lease would "most positively *not be extended*," which prompted more letters from Trenholm to Drayton. "I hope you are not dissatisfied with anything I have done," wrote the soon-to-be ex-miner. Referring to their shared capital account, Trenholm argued that his mining operation was less expensive than all others and "the debit to the acct has diminished very much." He hoped to begin paying regular dividends soon, which, he claimed, was "more than any individual, partnership, or company has yet done, or have any prospect of so doing." Despite more offers, Trenholm's appeals fell on deaf ears. Mathewes informed Drayton that Trenholm lost $4,776 during 1868. He mined Drayton Hall through 1869, and CMMC took over in January 1870. Frank Trenholm would never again run his own mining business.[75]

As Trenholm demonstrated, rock mining in 1868–69 was risky but potentially lucrative. Even Clifford and Mathewes admitted, in late 1868, that "no one as yet has realized any money in the working of phosphates." According to Holmes, CMMC had paid two dividends by late 1870, but as with much of the industry, the nature of those dividends is unclear. Did CMMC pay dividends from actual profits or pay out from unused capital, and were the dividends substantial or merely token? None of the other companies reported profits or dividends as of 1870. Former land-mining entrepreneur and state inspector of phosphates Otto A. Moses looked back to the industry's first years and summarized, "The tendency in mining is towards the laying of extensive works, and the purchase of large bodies of reserved territory. Small miners make heavy outlay in the beginning. . . . Some of them have been unsuccessful, and doubts have prevailed with capitalists concerning the security of investment in phosphate digging, but I know of no other species of mining where returns are as easily calculated."[76] Despite the risks, the rising number of local land-mining companies indicated that entrepreneurs believed that there was money to be made in the industry. CMMC's indirect takeover of Frank Trenholm's territory was an aberration. Less able than CMMC to take advantage of economies of scale, small producers nonetheless partook of the land-mining boom without fear of consolidation.

By 1870, phosphate land mining was well established as a substantial industry in the lowcountry. Revealing their worldview in promotional literature and articles on the companies, elite whites envisioned the industry as the savior of South Carolina and perhaps the South. Phosphate fertilizer could help restore the chronically worn-out soils of the "old" South and establish the long-delayed southern manufacturing base. To a landed gentry burdened by relative poverty and Radical Reconstruction, phosphate was a key to restoring the traditional race and class hierarchies. Upper-class whites could run the mines and the factories while fully utilizing their now-taxable lands and their cotton-exporting connections. Middle-class whites could supervise at the mines and factories. Lower-class whites could work in the factories, which in turn might attract immigrants to maintain a white majority. Finally, blacks could work the mines and help transport the rock.

For all whites, keeping the black majority busy (and thus harmless) during the slow agricultural periods ensured a degree of racial stability sorely lacking since emancipation. Not surprisingly, Holmes, Pratt, Shepard, Ravenel, and even Commins sought to claim the title of founder of the phosphate era and to pose as a community hero, a savior of the now "prostrate" state and section, and a father of a "new" South.[77] Unlike the founders, however, those who organized and ran the companies during the industry's first three years shared neither a common class nor region but merely the desire to make money and improve agriculture. In terms of business culture, South Carolina's land-rock-mining industry quickly grew beyond its founders and involved many of the lowcountry's leading planters and merchants, as well as northern fertilizer merchants and manufacturers.

Charleston did not enjoy a pro-business reputation before the Civil War. According to the *City Gazette*, the aristocracy had since the 1790s believed it to be "disreputable to attend to business of almost any kind." Hesitant to invest in what antebellum industry did exist, some of the city's elites deemed a planting, factoring, or legal career suitable for men of their class but seemed to frown upon investing in or managing an industrial enterprise as more the province of Yankees and Jews. South Carolina's preeminent antebellum leader, John C. Calhoun, promoted the fear that industry was a danger to the state's way of life. Even after the war, some members of the great Charleston families would only enter certain vocations. Unfortunately, some historians have read into this commercial snobbery a comprehensive economic "irrationality, entrepreneurial lethargy, or a pre-bourgeois mentalité" that pervaded the antebellum city and state and continued unabated into the postbellum period.[78]

The early phosphate entrepreneurs provide proof to contradict this assertion. After emancipation, the city's factors, shipping merchants, and lawyers realized that they too had been freed. Partially due to sympathetic newspaper articles, their enthusiasm spread throughout the lowcountry's planter elite. The initial refusal to back Nathaniel A. Pratt and Francis S. Holmes was more a factor of personality differences, strained finances, and scientific uncertainty than a general bias against business or industry. At the birth of the land-mining industry, Charleston's business elites and local planters, including Williams Middleton and George Trenholm, responded enthusiastically to the new industrial opportunities. The small group of entrepreneurs would rapidly expand in the ensuing years as phosphate mania took hold in the lowcountry.

Some historians have also claimed that the South became an economic colony of the North after the Civil War, but, again, South Carolina's land-rock-mining industry provides evidence to the contrary. Northerners did not overrun the lowcountry phosphate fields during the vulnerable 1867–70 years. Clearly, Philadelphia-based CMMC was the largest, best-financed, and most productive firm, but there was no trend toward, or conspiracy to create, a "colonial economy."[79] CMMC's superior funding allowed it to buy land and equipment in prodigious quantities, but Wando and the other local concerns also aggressively pursued land. Indeed, locals had the advantage in finding deposits outside the Ashley River basin, and Wando's dual role as a mining and manufacturing company weakens the colonial argument. Aside from first-mover CMMC, northern and southern entrepreneurs and their companies competed on an equal basis in terms of size of operations and output. Northern funding and management aided but did not dictate the terms of the early development of the mining industry. The remaining real regional difference was available capital, but this weakness would not affect most southerners in the land-mining and fertilizer-manufacturing industries for several decades.

Following the war, phosphate rock offered redemption to the lowcountry aristocracy. White men made the chemical discovery and organized the land-mining industry, companies, machines, transportation networks, and markets. None of them, however, intended to dig the rock themselves. Freedmen were the actual miners, digging the rock with shovels and picks and, later, working with steam shovels. Although in their writings most employers and industry chroniclers ignored or took for granted the land miners, labor was the most expensive, volatile, and important part of the equation. As in slavery days, black men clearly were not in charge and rarely able to voice their opinions. But just as they had in slavery, freedmen silently negotiated the terms of their labor, shaped the work to fit their needs, and took over much of the day-to-day working of the plantation. As Pratt predicted, phosphate land mining gave "employment to willing labor and bread to its hungry poor," but the poor laborer became a full partner to the scientist and entrepreneur in the development of the industry.[80]

3

Land Miners and Hand Mining, 1867–1884

Assumptions of mastery died slowly along the Ashley River. In the turbulent wake of emancipation, former slave owners such as Francis S. Holmes had to adapt to a world without a reliable, malleable, and abundant labor supply. Many of "their people" had fled the plantations, and former masters could not compel new arrivals or remaining laborers to work. The lowcountry rice industry had all but collapsed by 1865. Attempting to evolve from rice planter to phosphate-mining manager, Holmes complained that "laborers were scarce, and the negro, unaccustomed to such work, accomplished very little towards a day's task." He continued, "Where the Company expected to keep employed one thousand laborers, thirty could not be placed." For black laborers, the "new and untried field" of land-phosphate mining was too similar to rice work in the same fields under the same masters. Most refused to cultivate rice, and they were hesitant to dig phosphate.[1] An incomplete revolution, emancipation nevertheless empowered former slaves to demand new labor and social relations. The newly freed coerced white elites into negotiating rather than dictating. Unwavering in their desire for land and autonomy, lowcountry blacks used mining to advance their goals, but in doing so, they built up the industry and helped to liberate at least some of their former masters from the financial disaster of emancipation.

Freedpeople in the lowcountry shared with their brethren across the South common experiences after emancipation. Primarily, the ex-slaves wanted to replace planters' intrusiveness and control with their own autonomy and independence. To achieve these objectives, they resisted the restoration of overseers and gang labor, embraced education, and created

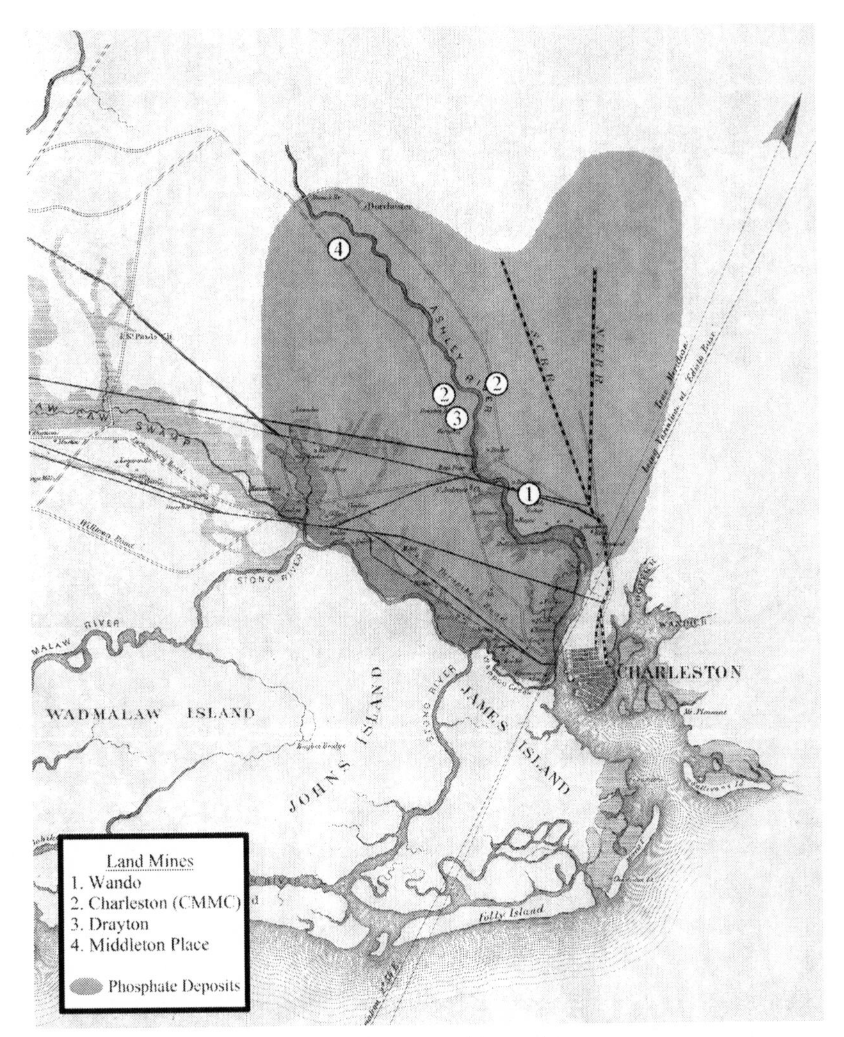

Map 2. Land-mining territories. Map created by author.

their own churches. Black families struggled against whites apprenticing their children, withdrew their women from the fields, moved around the lowcountry, and worked less. Throughout the Deep South, the postwar labor scarcity, together with favorable political winds, added to freedpeople's optimistic view that they could buy land and become independent farmers. With the failure of land redistribution, many reluctantly negotiated sharecropping agreements with white landowners. Bargaining from a position of relative strength with former paternalists, freedpeople, now able to leave

Figure 5. Paired hand miners, near Charleston, 1877. By permission of South Caroliniana Library, University of South Carolina, Columbia.

or threaten to leave obstinate employers, gained greater control over their lives.[2]

When land-phosphate mining began in 1867, the balance of power began to shift toward the men who did the physical work, the freedmen miners, and away from the factors, gentlemen-scientists, and entrepreneurs who did not. Although white owners and newspaper reporters seemed to take labor for granted, describing it as a cost or problem, like overburden or drainage, mining was hard work in an inhospitable environment. Laborers did not regard mining as a profession but, rather, as part-time work, and, like other post-emancipation occupations, a subordinate concern to enjoying their freedom. For managers like Holmes, the question in this new era was not "Who will dig?"—that the labor pool consisted of freedmen had never been in doubt—but, rather, "How will we get them to dig?"

As with almost everything else in the postbellum South, the labor struggle took place within the context of the past. South Carolina was Europe's leading supplier of rice in the eighteenth century, but its rice industry lost its primacy in the antebellum period, slightly recovered after the Civil War, and died early in the next century. The region remained, however, uniquely stamped by the rice culture, especially its task labor system.[3] Task dominated the language of work throughout the region, even for those who had never set foot in a rice field. Former slaves blended beneficial elements of the task system with the new privileges of freedom to create a work culture for mining, farming, and living in the lowcountry, and they forced former masters to accept it. Elements of continuity and discontinuity were fundamental to the working relationships of black miner and white manager, freedman and former slave owner. The struggle between laborer and supervisor over work time, pace, and conditions persisted after slavery, as did the basic power relationship between whites and blacks. But freedom created a more level field of negotiation. That discontinuity made the world seem foreign to elites and invisible to census takers, and it allowed the freedpeople to tailor mining to fit their new concepts of work and life.

Hand Mining

In order to understand labor relations in the phosphate fields, one must first visit those fields and appreciate the environment and labor itself. Hand mining took place in a region transformed by war and reclaimed by nature. Touring the Ashley River in 1871, editor D. H. Jacques exclaimed, "What a change!" Where once "magnificent cultivated estates and . . . fine mansions" lined the shores, there was "a land of ruins" and "a wilderness." He described phosphate-mining operations as "settlements," as if the region recently had shed two centuries of development. Nature was beginning to reclaim the slave-tamed countryside north and west of Charleston. In tangible ways, the not-so-grand planters and their now-free black laborers were starting over, resettling the region.[4]

The new wilderness was a formidable obstacle for miners. Rarely exceeding fifteen feet above sea level, the phosphate district was literally a "low country," predominantly swampy and bordered by river marshes.[5] Storm surges damaged wharves, boats, and washing machinery. While mining in trenches, workers battled rainfall, ground seepage, and tidal surges and often stood in ankle-deep water. In the warm months, malaria slowed or stopped work. The first miners hand bailed water but routinely endured the inconvenience. Later miners dug drainage ditches to keep the work area somewhat dry. By the 1880s, miners working for well-funded companies used portable steam pumps mounted on railroad cars.[6]

While battling water, miners cleared trees, underbrush, and overburden. Prospectors found outcroppings along the Ashley River and directed miners to remove a few feet of alluvial soil to reach the sixteen-inch phosphate layer. Early miners stopped digging when they reached the marl layer just below a tight grouping of phosphate rock, because the deeper rock, although often abundant, was not worth their time. Making up approximately 40 percent of the phosphate layer, the rocks were intermixed with pebbles, sand, clay, marl, or fossils. Generally white or off-white, the easily identifiable rocks appeared "rolled and water-worn," and "nodular," and they ranged from pea sized to a foot in diameter or more.[7]

Hand mining began as a haphazard venture, with inexperienced superintendents and miners sacrificing thoroughness for speed. Management and labor gradually improved methods of prospecting deposits, clearing trees and brush, organizing workers, digging rock, draining water, evacuating overburden, and hauling rock. The major advancements in mining occurred in organization, but technological and financial changes played important roles as well. The miners' primary focus, however, was improving working conditions and wages, and most work-related domains remained as contested as they had during slavery. The work was hard yet profitable, and they had some control over the work process.

While miners worked in large groups throughout the phosphate era, they also organized themselves into pairs. Pairing had been common in the Caribbean, where overseers assigned weaker slaves to stronger ones, and also in the antebellum lowcountry, where planters agreed to pair wood sawyers for weekly tasks. Working in pairs or family groups was part of the communal cooperation and flexibility that prevailed in tasking, and some postbellum miners chose to continue the practice. Although evidence of paired land-phosphate mining exists only in newspaper sketches and a DuBose Heyward novel, the practice likely persisted throughout the hand-mining era.[8]

The land-mining process began with landowners, entrepreneurs, scientists, and supervisors prospecting for phosphate deposits, but later, experienced miners became the day-to-day prospectors, with managers merely choosing promising fields each week or month. The industry's first prospectors did little more than find outcroppings that indicated shallow deposits, just as Holmes and Nathaniel A. Pratt had in 1867.[9] But outcroppings were a blessing soon exhausted, and prospectors turned to more systematic methods to find deeper phosphate strata. Holmes and Pratt's successors in the 1870s forced augers or pipes down until they felt solid resistance, which usually indicated the presence of phosphate rock. Workers bored for rock no deeper than seven or eight feet, the depth at which hand mining became unprofitable. They then dug test pits to determine the quality of the rock. In the 1880s and 1890s, the process became even more systematic, with surveys and BPL-quality analysis.[10]

Haphazard prospecting led to insufficient aboveground clearing at pioneer phosphate mines. Workers avoided live oaks and other large obstacles, leaving many fields partially cleared. Miners tended to dig their pits only where prospectors had found promising deposits. Hand miners and, often, superintendents preferred skimming the most readily available rock, rather than mining the company's lands thoroughly and systematically.[11] Whether mine owners accepted this imperfection as the price of the postemancipation labor market or condoned it under pressure for quick profits is unclear. Readily apparent was the fact that land mining and, later, river mining would continue to be beset by the competing goals of cost, speed, and thoroughness.

Derived from Holmes' crude marling operation, hand mining was costly, slow, and incomplete. Land miners used spades, pickaxes, and shovels to remove overburden and extract the rock from sand or clay.[12] Overburden amounts and phosphate rock sizes varied widely, even at the first mines near the Ashley River. At Feteressa mine, hands reached the rock after removing twelve inches of overburden, while at nearby Hard Farm mine, hands had to dig down four to five feet. Across the river, at Ashley Ferry, miners found phosphate rocks, ranging in size from an egg to a brick, eighteen to thirty-six inches below the surface. Miners then dislodged the rocks and heaved them into piles next to the pit. Digging phosphate rock was as onerous a burden as marling had been for Holmes' slaves and perhaps worse than most antebellum tasks. An observer described a skilled and "very industrious digger" expending great energy raising in "great masses—a dull, dirty stuff, all plastered over with black and slimy mud."[13] Enormous physical exertion in knee-deep water and mud in near-jungle conditions made hand mining an awful job. Not surprisingly, managers found it difficult to ensure that miners returned to work each day.

Miners toiled for mining companies as well as labor subcontractors. While many smaller firms supervised their own laborers, Charleston Mining and Manufacturing Company (CMMC) and other large companies often contracted for labor. Supervising seventy-five hands, subcontractor E. P. Toomer agreed to deliver to CMMC two thousand tons of rock per month during 1868. For entrepreneurs, hand mining was a comparatively inexpensive industry to enter, once they bought or leased the land. As shallow deposits (under eight feet deep) began to disappear in the 1880s and 1890s, companies gradually turned to steam shovels, thus raising the investment threshold and changing labor relations.[14]

Contemporary accounts yield a rough consensus of the nature and evolution of hand mining and how much it paid. Miners dug trenches, which became the norm in the late 1860s, in a parallel pattern across a field, in order to drain water, dispose of overburden, cross the field, and maintain level ground for rock transport. They insisted on payment by the task, which, through negotiations with supervisors, became a twelve- to fifteen-foot-long and six-foot-wide trench. By the 1880s, the going rate was 25 to 30 cents per foot of rock and overburden, so most miners earned $1.00–1.20 each day, over twice the wages of South Carolina's farm laborers. Other tasks, involving transportation and processing, had their own pay scales.[15] Hand mining's task size continued into the steam-shovel era. With machines digging trenches measuring fifteen by five hundred feet, laborers stationed at six-foot intervals separated phosphate rock from overburden. Working with steam shovels in 1892, the miners had the same daily task and earned the same wage per foot as they had a decade earlier. As late as 1914, hand miners worked by task and alongside the machines. Although hand mining persisted into the twentieth century, the advent of steam shovels took much of the skill and control out of miners' hands. Before the 1890s, however, land-phosphate miners shaped their environment and, to a certain extent, controlled the destiny of the industry.[16]

Whether working only by hand or with steam shovels, miners had to dispose of overburden and transport the rock. Laborers tossed overburden into the closest mined trench and piled the rock on the other side for transportation to riverside washers. Mining in the late 1860s, hands initially looked to Holmes' marling precedents by loading the rock into wheelbarrows or small carts and zigzagging around mined or filled trenches to the washers. Better-funded companies erected narrow-gauge rails on which workers or mules moved tramcars. The rails penetrated the mining field, but wheelbarrows still carried the rock from trench to rail.[17]

As miners exhausted shallow deposits near the river and distances from field to washer increased, well-funded companies like CMMC bought steam locomotives to move tramcars and to keep most of their workers mining instead of pushing wheelbarrows. Companies without locomotives suffered from transportation bottlenecks. With twelve miners, eight "barrow wheelers," and five mule car drivers in October 1869, Williams Middleton's Ashley Mining & Phosphate Company of South Carolina (Ashley Mining) could not move the rock over a mile from trench to washer fast enough and at times had to divert all miners to the activity. Locomotives became more common in the 1880s as distances to the washer increased to over two miles.[18]

CMMC was a leader in making mining and hauling more methodical. In 1882, state phosphate inspector Otto A. Moses described the company's workers laying out a main trunk line with laterals six hundred feet apart and a "line ditch" splitting the distance between laterals. Two sets of miners began digging at right angles from the laterals, with each man responsible for several consecutive pits measuring twelve by six feet. The result minimized wheelbarrowing, maximized locomotive use, and more thoroughly depleted the field. Foremen "sharply scrutinized" each full wheelbarrow for excess overburden as workers loaded rock onto tramcars. Moses admitted, however, that supervisors and miners often altered the system in swampy areas or former rice fields, reverting to "the single pit system."[19]

The amount mined and the destruction generated was impressive. Hand miners during 1868–70 removed about six hundred tons of phosphate rock and 3,000–4,500 tons of overburden per acre. CMMC's laborers mined about 150 tons of rock per day in July 1868, and daily averages increased to seven hundred tons or more in the early 1880s. When steam shovels and hand miners shared deeper trenches, the average per acre increased to about 700–800 tons of rock and more than six thousand tons of overburden.

After mining, the field looked to observers "as though a whirlwind had passed over it." Trees and undergrowth grew back quickly in the semitropical environment, but the trenches and overburden mounds remained for decades.[20]

Task System

Before they dug phosphate trenches, most early miners had been slaves on lowcountry rice or mixed-crop plantations under the task system. Firmly established prior to the Civil War, tasking is the key to understanding the otherwise murky negotiations between miner and manager, freedman and former master. Rice and task made lowcountry slavery fundamentally different from slavery elsewhere. Most American slaves on large plantations labored all day in gangs, "driven" by white overseers or black "drivers," with little control of work pace, few moments to relax, and no "free" time until after sundown. Lowcountry slaves, by contrast, worked on tasks, not in gangs, and received incentives, not threats, to work harder. Overseers or drivers assigned slaves tasks, or specific daily jobs, and then supervised in a perfunctory manner. Ironically, rice cultivation demanded the largest and most well-coordinated group of slaves, and yet task slaves had more autonomy than gang slaves. Slaves had the rest of the day free after they performed their task, and this time became so "sacrosanct" that they expected payment when asked to work longer.[21]

Two antebellum hybrids of the task system likely reappeared in phosphate mines. The first one involved squads of fewer than seven slaves working at scattered sites where close supervision was costly or complicated. The second hybrid, collective tasking, featured larger groups completing tasks. Together with individual tasking, the two variants allowed planters, overseers, and drivers to stay out of the disease-ridden rice swamps and to reduce the costs of supervision while maintaining an "industrious, efficient, manageable, and contented" workforce. And slaves, seeking to shape a more palatable form of servitude, negotiated a degree of control over their work and social lives.[22]

Comparisons of several aspects of rice and task slavery and phosphate mining illuminate a significant amount of continuity between the two eras. Rice and rock, unlike sugar and tobacco, were hardy crops that involved no procedural synchronization and little supervision. The task system's reliance on slaves' daily expertise made the transition to phosphate trenches more successful. Absentee rice planters hired black slave drivers to act as overseers, and drivers, or foremen, supervised mines as late as the mid-1880s. Whether they were effective, however, given the miners' autonomy and growing proficiency, remains unclear. Digging and water-control tasks on antebellum rice plantations were easily convertible into postbellum mining jobs.[23] Finally, the basic unit for a ditching task on a rice plantation was the "compass," a five- by 150-foot section of the field on a sixtieth of an acre. Although shallower than a phosphate trench, the rice compass was of a shape that suggests its continuation in the mining era.[24]

Although it had advantages for master and slave and, later, for miner and manager, the task system was not free of conflict. Workers challenged any increase in task size and fought for the right to define themselves as full, half, or lesser hands.[25] For slaves and miners, the appeal of the task system lay in the autonomy it fostered for work and family life. Less supervision increased the distance between management and labor and decreased the potential for friction during and after work. The slave and miner had a degree of control over work "pace, rhythms, and routines" and the length of the workday. After completing the task, slaves worked for overtime payments or hunted, fished, raised livestock, and grew crops. They created an "internal economy" within the lowcountry, selling to their masters or nearby slaves. Miners also worked partial days, took many days off, and extended the internal economy. Managers' efforts to curtail this expanded autonomy were largely unsuccessful.[26]

Historians are divided on the impact of antebellum and postbellum tasking. William Dusinberre admits that the task system was widespread in the region, but in emphasizing the horrors of lowcountry slavery, he de-emphasizes the slaves' autonomy and internal economy.[27] John Scott Strickland argues that lowcountry blacks rejected the "emergent capitalism of the postwar South," and while accumulating some property, fought to secure independent subsistence. In contrast, Philip D. Morgan more properly underscores the powerful element of entrepreneurialism in post-task home production.[28] Larry E. Hudson Jr. maintains that lowcountry slaves leveraged the task system to become better off than gang slaves, in terms of social autonomy, economic opportunities, and the chance to create a relatively fulfilling life, all of which benefited them during the transition to freedom.[29] To this historian, the after-task work of slaves and miners seemed to combine a subsistence lifestyle with sustained efforts to earn money or goods, accumulate property, and get ahead. Both subsistence and cash production served to insulate slaves and freedpeople from the double-dealing and controlling tendencies of area whites.

Emancipation

With emancipation, the lowcountry's freedpeople sought independence and autonomy by protecting parts of the task system and trying to accumulate property, especially land. They chose from among several sharing arrangements, including the share wage, share rental, and two-day systems. Sharing the land and risk, they saw themselves as partners with, rather than employees of, landowners. Black agricultural workers used the postwar labor shortage as an effective bargaining tool, threatening to abandon the fields of inflexible planters for better deals elsewhere. Control of their labor meant that former slaves created autonomous social lives as well. Sharing systems generated space between black workers and white landowners.[30]

The lowcountry's newly emancipated workers, many of whom would become phosphate miners, had well-established ideas of what freedom meant, including continuing the favorable and adjusting the less-favorable elements of the task system.[31] Former prime hands insisted during 1865 and 1866 that they work as part-time, fractional hands and that contracts include payment with wages or crop shares, both on the basis of tasks completed. Freedpeople often refused to do postharvest rice work, especially ditching and draining, because they suspected that planters would try to change it into gang work paid monthly. They also feared that such labor invested in next year's crop would cede control over future production to planters. Freedmen justified their demands more in terms of free time than working conditions. In significant ways, then, emancipation was less of a revolution in the lowcountry than elsewhere. While freedom empowered most American slaves to transform their work, the "overriding imperative" of lowcountry freedpeople was to avoid any fundamental changes to the structure of their workday.[32]

Freedpeople had deeper reasons for demanding continuation of the task system. Born and raised in the lowcountry, they had a strong sense of place and now free, expected the land on which they had so long toiled to become theirs. When their hopes for "forty acres and a mule" faded early in Reconstruction, they fell back upon their former, more limited, claim to the land, the "right" to a plot of land after they completed the task.[33] This devotion to tasking's benefits spread to the many freedpeople who relocated to the lowcountry. Even those used to gang labor, when acquainted with the advantages of the task system, adopted and defended it as their own.

Mining managers had little choice but to put up with freedpeople's demands, because phosphate mining was so similar to the labor slaves performed but hated on lowcountry rice plantations before emancipation. Miners worked in identical conditions on many of the same rice plantations with the same tools moving what slaves had considered nuisance rocks. Rice ditchmen, who built dikes and dug drainage ditches, had similar daily assignments as miners. Slaves who had built and maintained rice-threshing mills may have noticed a superficial resemblance to phosphate "mills" (washers).[34] Not surprisingly, then, managers converting rice tasks into mining tasks made substantial concessions to miners.

Lowcountry agricultural laborers adapted the task system to the post-emancipation world by negotiating the two-day system, a unique form of sharecropping that created time for mining and other pursuits. Cultivating various crops in someone else's fields two or three days each week and supplied with land and lodging, freedmen on their off days left the plantation to find other work while their wives tended crops and children. Freedpeople also bought, sold, and bartered in the neighborhood-market economy, thereby decreasing white interference or control. Although the two-day system modified the task system by measuring contractual obligations in terms of time rather than tasks, it continued tasking's benefits of personal time and access to land. Whites disapproved of this unprecedented level of black autonomy. Planters protested that freedmen rarely worked a "full" day, Freedmen's Bureau agents complained that it promoted subsistence farming, and northern reformers preached the Protestant work ethic, but all to no avail. The two-day system and its variants became entrenched in the lowcountry a few years after emancipation.[35]

Land-phosphate mining became an important part of freedpeople's plans within the two-day system. Mining companies, earning money from each rock shipment, offered miners short-term cash wages, unlike planters who paid out at the end of a growing season. Contracts, shares, and liens were not necessary in the mining industry. Freedmen could travel several miles to a mine, work one or two days, and return in time to tend their land and work for the planter. Most mining jobs required little skill, so managers generally hired any strong hand that showed up. While supervisors complained about the widely varying numbers of miners, the flexibility suited miners and their families.[36] As a good source of cash and an alternative to agricultural labor, readily available mining jobs strengthened the negotiating position of lowcountry freedpeople.

Mining also complemented the planters' agricultural calendar. During the summer lay-by (between the planting and harvesting times) and in the winter months, contracted freedmen had fewer obligations in the fields and therefore more time to travel to distant phosphate mines, leaving their families for extended periods to live in company-owned dormitories. Planters appreciated seasonal shifts to land mining, because the industry absorbed their temporarily idle black laborers while not permanently diverting the labor from the fields. In addition, "supplementary short-term labor" such as phosphate mining kept a pool of other temporary, noncontracted workers in the area for busy planting or harvesting seasons.[37] Mining offered flexibility for both manager and laborer, but in practice it especially favored lowcountry laborers who had family support and access to land.

1870 Census

While mining became a vital economic activity in the lowcountry, the 1870 federal census recorded only a fraction of that activity. It did illuminate much about the miners in terms of race, housing, and marital status, as well as freedpeople's economic strategies. Unfortunately, census enumerators vastly undercounted phosphate miners, because they failed to translate the freedpeople's agricultural goals, multiple occupations, and domestic arrangements into meaningful statistical categories. Faced with former slaves who farmed, mined, hunted, and fished during various times of the year, census takers recorded the agricultural job and ignored all other economic

Figure 6. Workers and "quarters," Pacific Guano, Chisolm's Island, circa 1889–95. Courtesy of Beaufort District Collection, Beaufort County Library, Beaufort, South Carolina.

activities. Seasonal migration and fluctuating housing arrangements similarly caused land miners to fall between the wide statistical cracks. The result obscured the importance of the industry and its workers in census and, consequently, historical publications.

Census records indicated that most miners were black, illiterate, adult, and native to the state. In the most populous mining district, St. James Parish Goose Creek, only one of the 252 black phosphate workers could read or write, and none was born out of state. Miners varied in age from twelve to sixty-one, and the average age was thirty. That the twenties was the most populous age category indicated that the job was physically demanding. They were older workers within a younger population.[38] Their demographics, coupled with the region's rising population, suggest that many miners had been rural slaves who migrated nearer to Charleston after emancipation.

Despite their mature ages, the miners listed in the census did not live in family settings. Less than 3 percent of St. James Parish Goose Creek's black miners likely lived with their nuclear families, and the vast majority lived in single-sex group housing. Several companies provided dormitories near the mines to improve worker turnout. Ashley Mining supplied housing by May 1868 while CMMC had "houses and sheds" in 1870. Palmetto Mining and Manufacturing Company built "comfortable cabins for the negroes, white cottages, [and] a country 'store'" soon after.[39] A seasonal solution for some workers and a more permanent option for others, company housing (and stores) became an industry standard by the 1870s.

Group housing provided census workers with the most easily identifiable group of phosphate miners. Of those recorded in the census, 92 percent of the black miners lived in rural areas in mostly single-sex groups of nine to twenty-eight. Many in each dwelling shared the same surnames, suggesting that male family members worked together. For example, Abraham, Jack, and William Ramsey shared a group housing unit in rural Charleston County with John and Henry Anderson and Henry and Peter Adams. Of the few white miners, all lived in the city with other whites or their families in hotels and boardinghouses. Only one white man, New Yorker A. Gindusteau, lived in group housing.[40] Census inadequacies aside, the substantial number of black miners living in group housing reflected the still-unstable state of labor in the postwar lowcountry, especially the influx of rural single men.

Most miners included in the census, as well as those overlooked, fell into three life circumstances in 1870. The first group included rootless men, single miners without family who had moved from upstate and now mined year-round. These refugees congregated near Charleston after the war, and many found work and shelter at the mines. Group housing was a somewhat permanent situation interrupted only by occasional excursions to the city, odd jobs in the area, or periods of economic inactivity. Their living arrangements reflected a comparatively unstable society lacking immediate family, church and community groups, landownership, and property accumulation. Men in the second category supported their families by seasonally migrating to the mines, perhaps dozens of miles away. For these family-oriented migrants, the temporary shift into group housing was less disruptive (than the first group's experience), because mining allowed the men to earn higher and more immediate wages than farm labor and then return to subsistence and sharecropping nearer to home.[41] Freedmen in the third group lived with families nearby and worked mine and farm during the week. Only these men could collect the higher wages and continue their home life. The first group, often living in group housing, was an easy target for census enumerators, whereas the second group's census listing depended largely on when in the agricultural calendar the enumeration took place. Census workers often overlooked the third group, because the men alternated between subsistence, agricultural, and industrial economies.

No records of freedmen negotiating wages, conditions, or work pace with mine owners or superintendents exist, but negotiations of some sort surely took place, as they did during slavery, and men of the third group enjoyed advantages over the others. The ability to withdraw their labor, create a measure of subsistence, and be mobile throughout the year gave them bargaining strength with the former slave owners who owned or managed the mines. Freedmen living within a radius of a dozen miles of the mines could choose when to work their own land, when to visit Charleston, and when to work for wages, either as agricultural laborers or phosphate miners. The subsistence freedman was less prone to be bullied, cheated, and abused at the mine and could leave when conditions proved unsatisfactory.

The 1870 census snapshot left many questions unanswered regarding the three groups of phosphate miners. How long did men stay in the camps, and how frequently did they travel to the city? Was it possible to live in group housing and meet marriage partners or establish what was then considered a stable life? Were they dependent on company stores, or did family members keep them supplied during their mining tour? How often did they mix mining with the two-day system? How far did they commute? Did the mining wages enable wives to withdraw from the fields? Were they recruited by companies or part of community networks? Did they have garden plots or acreage to plant? Did they save any wages and eventually buy land? To what extent were they in debt to local merchants or liquor dealers? Unfortunately, the answers to these and many other questions remain hidden from census and other records.[42]

Literacy statistics help to fill in some gaps. Born around 1840, many miners had grown up as slaves and were forbidden to read. Although roughly 5 percent of slaves circumvented the ban, those in rural areas had fewer opportunities to achieve literacy than did their urban counterparts, a trend that continued after emancipation. Schools such as Charleston's Avery Normal Institute and St. Helena Island's Penn School ably served lowcountry blacks, but few schools existed in most rural areas, especially in the mining districts north and west of Charleston. Over 99 percent of black miners counted in St. James Parish Goose Creek were illiterate. Thus, phosphate-mining regions had less in common with black Charleston (45.4 percent literate) in 1870 and more in common with the rest of black South Carolina (19 percent). The desire of lowcountry freedpeople to learn was omnipresent, but the time and money to attend school often was not, especially for rural young men supporting families. Struggling to make a living and with their wives withdrawn from field work, those living outside the city were hard-pressed to justify school over work. For freedmen in St. James Parish Goose Creek, the constant demand for labor in the phosphate mines elevated the higher and more tangible priority of cash over education.[43]

Freedpeople had clear priorities and diverse strategies, but census enumerators did not understand them. The result was the omission of land mining in census compendiums and the discounting of its impact on the state's economy and society. Contradicting census reports, contemporary newspaper and industry accounts indicated that the industry was much larger. The disparity lay in the nature of seasonal, part-time work in the lowcountry.

The extent of, and reasons for, the 1870 census undercount are evident in the industry and populations schedules. The "Average number of hands employed" totaled 356 in the industry schedule, but enumerators ignored at least fourteen mining companies, including the dominant CMMC, thus rendering the total virtually meaningless. Extrapolating yield-per-hand census data to the unlisted companies suggests a total of 968 employees. The population schedule listed an even less accurate 290 men, but it did reveal that black men held the more menial jobs and whites the managerial or sales jobs. Few whites and women (aside from cooks at some dwellings) labored in the phosphate fields.[44]

Contemporary newspapers provide additional evidence that the census undercount was substantial. The *Charleston Daily Courier* reported that during the first full year of production, the industry employed "many hundreds of persons" and that CMMC paid 130 hands in July 1868. By February 1869, the newspaper reported that "several thousand 'freedmen'" worked the mines. The city's papers also testified that the industry's rate of production grew 160 percent from 1868 to 1869 and 104 percent from 1869 to 1870, signaling a proportionate rise in employment.[45] While not irrefutable, and discounting its boosterism, the newspaper evidence adds weight to the conclusion that census data was substantially inaccurate and that the estimate of 968 functions as a more realistic minimum.

Why, then, did census enumerators miss most of South Carolina's land-phosphate miners? Although the 1870 census was especially flawed in the politically chaotic South, the undercount was sizable even by 1870 standards and was directly related to peculiarities of the industry and its workers.[46] The problem originated when census enumerators asked low-country freedpeople for their one occupation. Working several jobs within the agricultural calendar and their economic strategies, freedpeople hoped to become independent farmers and viewed mining as a part-time job, not a career. Undoubtedly, many of the thousands listed in the census as "hand" and (farm) "laborer" mined phosphate at some time during 1870. But unless they were living in company-built group housing and actively mining, they were unlikely to volunteer "phosphate miner" as their one occupation. Independent farming was their goal, and short of that, most identified themselves as agricultural workers.

The undercount also originated in the clash of Victorian and African American assumptions regarding the family and in federal and planter reactions to postwar labor unrest. Burdened by gendered assumptions, the Freedmen's Bureau promoted freedmen as "self-reliant and manly" and placed freedwomen squarely in a separate domestic sphere. Bureau agents pressured freedmen to honor their marriage "contracts" by providing for the family and rationalized freedwomen's field labor as a temporary expedient. The Bureau's assumptions ignored and conflicted with the values of lowcountry African American families. Appreciating the obligations of the extended family, the economic contributions of children, and the material basis of marriage, freedpeople sought to control the terms of wage labor and the organization of the family economy and to keep planter and overseer interference to a minimum. Men's work was one part of a strategy, not the family's sole support, and did not necessarily mean one job during the year.[47]

Another source of census confusion was the precarious nature of plantation housing for freedpeople. After the war, the Bureau protected freed families' access to housing, even when one relative incurred the planter's wrath by working off the plantation. With lodging and access to land, black families had their basic necessities covered while their men negotiated better opportunities off the plantation. By 1866, planter complaints about what they termed labor unrest and agricultural inefficiency led the Bureau to support planters in evicting entire families when one of these "troublemakers" obviously worked elsewhere. The change in policy forced freed families to be more discreet.[48] By keeping a low profile, freedmen could still house their families, have access to land, and pursue more lucrative jobs, including phosphate mining. Miners continued to blur the line between plantation laborer and "troublemaker" after the Bureau retreated from South Carolina in 1869.[49] Not surprisingly, when interviewed by federal census workers, most laborers posed as farmhands, not phosphate miners, in order to avoid eviction. Only those living in mine housing felt free to sever the link to farming.

Adding to the census morass, enumerators' interviews and interpretations yielded a variety of job titles that baffled census statisticians and historians. The population schedule from St. James Parish Goose Creek included five categories describing the same job ("phosphate laborer," "phosphate miner," "phosphate," "phosphate digger," and "digs phosphate"), likely diluting the perceived size of mining. In addition, mining districts in Charleston and Colleton Counties had numerous listings of "laborer" or "hand," ambiguous categories undoubtedly including farmworkers but probably miners as well. Unsure how to categorize freedmen with multiple economic pursuits, enumerators often lumped them all into a general "laborer" category. Even the word "phosphate" created confusion. Charlestonians commonly referred to manufactured fertilizer, land rock, and river rock as "phosphate" and used the terms "phosphate" and "fertilizer" interchangeably. Puzzled enumerators then classified many miners working for dual-purpose companies (mining and manufacture) as factory operatives. The result was the simultaneous listing in census summaries of 316 fertilizer operatives and 2,501 fertilizer hands in the United States and of 825 fertilizer hands in South Carolina.[50] Statistically, then, many miners simply disappeared.

Timing was a major factor in the undercount, because census takers surveyed Charleston County's districts from June through November

1870, and statistics in mining areas varied widely depending on the agricultural calendar. After planting season, many landowners often dismissed freedpeople, leaving them to grow food or work elsewhere. Mining regions surveyed before and after the summer lay-by (mid-June to late August) registered mostly farm laborers, while surveys during the lay-by revealed a majority of miners. This seasonal effect was evident as early as 1868, when CMMC's workforce numbered about 100–130 men in July but declined to 75 in September. In 1870, enumerators found 257 phosphate workers in St. James Parish Goose Creek but only during July and August and only in group housing. Further complicating the picture, census data from other Charleston County parishes (with no mining) revealed that some miners commuted up to a dozen miles to mining regions.[51]

In 1870, lowcountry laborers were mobile and creative in their approaches to economic security and life strategies and often preferred land mining's wages to those of farming. The miners were a blurred mob in the eyes of a white, formerly slave-owning society and moving targets to white census takers. The census undercount helped to obscure the industry's historical legacy, because successive generations ignored industries not appearing in census summaries. Based on an estimated 968 hands, land mining was the state's second-largest industry in 1870, trailing cotton manufacturing by less than two hundred employees. But due to the occupational and seasonal confusion in the census population schedules, phosphate mining did not appear in summaries of the federal census, despite sharing similarities to other South Carolina "manufactures," most notably fertilizer, lumber, and turpentine.[52]

Middleton Place

Eric Foner describes freedom during Reconstruction as a "terrain of conflict" and explains that "between the planters' need for a disciplined labor force and the freedmen's quest for autonomy, conflict was inevitable." But James C. Scott argues that the various "weapons of the weak" rarely included direct confrontation. A close look at Middleton Place's mines illustrates freedpeople's strategies and exposes managers' responses during the years before the 1870 census. Problems with workers, finances, and production surfaced almost immediately due to miners' assertiveness,

Williams Middleton's management style, and his company's parsimonious investors.[53] An experienced if not hands-on former slave owner who owned several rice plantations, Middleton attempted to resurrect rice cultivation and control the miners, but he fought the legacies of task labor. He soon shed his paternalist ethos and insisted that the freedpeople mine phosphate and cultivate rice during a "full" work week, sunrise to sunset for six days. He increased wages and provided stores and housing, but he found that retaining miners was a daily challenge. Used to mastery over "his" people, the middle-aged master struggled to adjust to altered postbellum realities.[54]

Middleton Place was near the center of the Ashley River phosphate district. Following years of postwar financial agony, Middleton directed laborers to begin mining in February 1868, and by April, he had come to terms with Baltimore businessmen Charles Baker and Robert Turner in establishing Ashley Mining.[55] As the mine's manager, Middleton assembled desperate freedpeople to mine, but their desperation only added to his burdens. He complained that they were "so destitute" the company had to supply food, mining tools, and "shanties," but he made them pay for the supplies with their labor. He also provided "commissary stores" and housing for the workers. Temporary measures, the shanties were for summer use, at the mines, and for miners only. Besides equipping miners, Middleton built a tram road and wharf to move the rock effectively. He protested to his partners in May 1868 that "we cannot get on . . . managing in this manner" and that "we have to resort to so many expedients that we cannot yet work economically."[56]

Overwhelmed with start-up costs and familiar with slaves' proclivity to break tools (to delay work), Middleton initially demanded that freedmen buy their own mining tools. In theory, this would save the company money, thwart work stoppages, and induce more work. This policy followed a similar trend with lowcountry planters who made tool ownership part of the two-day system. Unfortunately for Middleton, shovels and picks worked equally as well at other mines or on freedpeople's garden patches and therefore proved to be poor incentives to regularly mine Middleton Place. By early 1871, Middleton had capitulated, and the company, not the miners, now owned the tools.[57]

Middleton also complained that Ashley Mining had to compete with the Freedmen's Bureau for laborers. Despite ample evidence that the federal agency favored planters' interests, Middleton believed that "the 'Bureau' [is] feeding every idle vagabond from our neighborhood," luring them into Charleston, and making it more difficult to find familiar hands in the "country." Middleton need not have worried. The presence of the Bureau in South Carolina peaked during mid-1867 and declined precipitously until 1869, when its role effectively ended.[58] Despite protests by Middleton and others, the agency had an insignificant impact on the supply of phosphate miners.

Middleton lashed out at the Freedmen's Bureau, but the real threat was the freedmen's mobility. The lure of movement was strong after the restrictions of slavery. Seeking land, education, relatives, society, economic opportunity, and physical security, upcountry and lowcountry freedpeople moved nearer to coastal cities. Charleston's black population rose from about 42 percent of the total population in 1860 to 54 percent a decade later. In the fall of 1865, a visitor commented that the city was "full of country negroes," an observation likely not lost on local whites. The lowcountry had a black majority (about 68 percent) between 1860 and 1880, and that majority increased throughout the period in Charleston and Beaufort Counties, mainly due to their namesake cities.[59] Although part of Middleton Place was in Charleston County, the distance to the city was fourteen miles, too far for a miner's daily commute but close enough for a miner to spend days or weeks in the city and to move there permanently when better opportunities arose. The rest of Middleton Place was in Colleton County, a district with no substantial city and a declining black population.[60] Ashley Mining's demographic disadvantage led to frequent labor shortages.

The proximity of cities strengthened ex-slaves' weapons, mobility and subsistence, and helped to create the postwar phenomenon of "walking away." Hardly the Republican Party's ideal, free labor in the South meant that blacks proclaimed their freedom and bargained with employers by leaving work and staying away indefinitely. The tactic was effective in a tight labor market where freedpeople subsisted on garden plots and by hunting and fishing. Middleton detested the free labor "experiment," bemoaning the "uncertainty of negro labour" and miners' tendencies to "go & come at their own pleasure regardless often of the sacrifice of wages."[61] Like many ex-slaveholders, he instinctively blamed the free labor system for black assertiveness and his lack of control over labor. Complaining to his partner about a recent hot spell during which miners fainted, Middleton reasoned that "had I been assured of my ability to collect them again, I should have discharged them for a few days," but he feared such compassion to be "too dangerous under the circumstances."[62] Miners' walking away caused "an infinity of trouble" for Middleton: "So much depends upon negro caprice it is difficult to feel sure about anything. The hands break off upon every imaginable pretext. They do little or nothing before 12 o'clock on Monday, and never do anything after 12 o'clock on Saturday. [They are] all crazy upon the subject of 'going to farming.' Poor wretches!"[63]

Nearby mining companies had similar labor problems, indicating that freedpeople may have been playing Ashley Mining and the other firms against each other. In September 1868, Middleton noted that two nearby mining companies had shut down to "reorganize" using white labor. Amid the increased scarcity of agricultural laborers after emancipation, southern planters reversed their stand against white immigration, and white South Carolinians led the new drive to replace what they termed the "lazy, shiftless, and uppity" black workforce. Earlier in 1868, Middleton considered hiring Irish labor, but he continued with black labor out of necessity. Seeking mastership over labor, southern planters should not have been surprised by how few immigrants took up their offers.[64]

Middleton's biggest difficulty was the combination of seasonal labor fluctuations, laborer mobility, and freedmen's memories of slavery. His workforce grew from 50–60 miners in mid-June 1868 to 100 in late July and to 80–150 in late August and then declined to 50–60 in mid-September. Despite his understanding of the agricultural cycle, Middleton was dismayed by the dramatic fluctuations in available miners. He was not used to "his" laborers walking away to tend others' crops, including their own. Clearly however, more was at work than just seasonal variations. Freedmen rejected full-time mining at Middleton Place, especially mining near the rice marshes, because it resembled "times of slavery." Middleton's insistence that the former slaves mine a relatively full day and also cultivate rice only underscored such negative comparisons, and they vigorously resisted his efforts to form gangs.[65]

Middleton sought to outflank the miners' walking away first with higher wages and then company housing. New to the free labor market, he tried a variety of pay schemes, including wages by the hour, day, ton, and barrowful.

With the cost of digging about $1.50 to $2.00 per ton in early September 1868, Middleton paid miners what he considered a good wage, 75 cents per day. Within weeks, however, he reported to investors that "strikes" along the Ashley River forced him to raise wages. By early October, his workers demanded an increase from 10 cents to 12 cents per barrowful.[66] Chronically short of cash and competing for labor with the well-funded CMMC, Middleton eventually learned that bribing miners to work paid diminishing returns.

Providing company-owned dwellings for reliable labor was more effective than bribes, because it benefited both miners and management. Rootless migrants to the lowcountry needed affordable housing relatively close to Charleston. For freedmen living farther from the mines than a daily commute, company housing made mining a good option during the lay-by and other slow times. Newly built housing was cost effective for the company, because it was a reusable way to keep miners on the job. Middleton gained some labor control by restricting the houses to "dependable" workers and saved additional cash by likely using outlying slave cabins. The housing program began in May 1868 when he directed miners to build shanties, lightweight and makeshift structures for summer shelter. As mining moved farther from the river by early September, Middleton supervised permanent improvements, including "pineland houses" closer to the current diggings.[67]

Labor troubles in September 1868 and success in boarding reliable black labor led Middleton to plan better houses to attract white miners. He stressed to Baker that housing white workers at the "summer settlements" (presumably with black miners) was not an option. The Baltimorean agreed with the plan and offered to send "experienced miners," likely white native-born workers from the Northeast. Middleton also sought out a Mr. Coxe from Philadelphia to recruit skilled coal miners. Hardly a shanty or slave cabin, the white miners' house was to cost about one thousand dollars, be two stories high, and include nine-foot ceilings, glazed windows, and plastered walls. But Middleton failed to lure even inexperienced whites to mine phosphate. Compounding this failure, mining continued to move farther inland, away from existing company houses and slave cabins. By late November, Middleton needed to build six "course houses" at a cost of eighty-six dollars each to "keep the hands together in cold weather."[68] Middleton was dependent on freedmen and recognized their need for housing, but his housing policy displayed a persistence of antebellum notions that likely contributed to Ashley Mining's labor problems.

Building company houses led Middleton to establishing a commissary, which he had planned as early as 1866 for rice workers. After emancipation, many planters, including his cousin Ralph Izard Middleton, built plantation stores to offset declining agricultural income, induce freedpeople to work, and recapture some of their wages. While Williams Middleton initially complained that the store was yet another expedient straining company finances, by June 1868 he realized that he could inflate the prices and still attract miners, so he began to plan a larger store. The stores gave Middleton—more important than profits—some control, by supplying necessities and through debt, over his labor force. For freedpeople, previously unavailable products and liberal credit made the stores irresistible attractions.[69] However, they were also terrains of conflict. Just as they did as slaves, freedpeople fought back by plundering the commissaries. Middleton bitterly complained on Christmas 1868 that the stores had been robbed twice within the month and were not profitable. The following October, he mentioned the need for "utmost vigilance" in protecting rice supplies. But he did not close the stores and even converted a "rice house" into another mining commissary. Resigned to bargaining with freedpeople, Middleton wrote Baker that the stores were a necessity "without which we cannot carry on the company's business."[70]

Middleton's miners walked away and stole rice, but they did not strike. When he wrote to Baker about strikes, he meant that the miners had "walked off without any warning," implicitly demanding higher wages. Like freedmen across the South, the miners negotiated with slave tactics. Protesting as individuals, staying nearby, and avoiding direct confrontation, slaves neither challenged white authority nor endangered white security. Masters tolerated such "truancy" and doled out light punishments. Freedmen adapted the effective strategy, wielding labor mobility as a weapon against poor wages and conditions. Scott argues that such "weapons of the weak" were most effective for the world's subordinate classes, because open, collective, and organized resistance was "dangerous if not suicidal."[71] Although they had gained more control of their working and personal lives since the end of the war, freedmen were "weak" in the sense that they were black in the white-dominated South and did not own the phosphate land or the companies.

Unable to fully exploit his former slaves and satisfy northern investors, Middleton quit the phosphate-mining business after three difficult years and leased his land to other entrepreneurs. He failed as a mining manager because he did not understand the contours of the lowcountry's new labor landscape. Ignoring freedpeople's preferences within the task system, Middleton attempted to create a mining plantation based on full-time wage labor. As a result, he faced almost daily labor shortages. The Charleston native was no more ready to negotiate with the freedmen than were the Yankees who came after the war expecting to implant their version of free labor.[72] When Middleton tried to keep freedmen near the mines by raising wages and building stores and houses, he was merely bribing the most desperate workers, not addressing their fundamental priorities. And, in doing so, he underestimated the freedpeople's desire and ability to use the industry to shape their meaning of freedom.

1880 Census

The 1880 census was an improvement over the previous edition, but systematic errors remained. Data on land miners' wives and children was more revealing, and all enumerators tallied in the same month, thereby reducing seasonal variations. Most fundamentally, the census exposed a dramatic shift in the miners' domestic circumstances. Unfortunately, the size of undercount increased due to continuing enumerator errors and terminology problems. As with the 1870 version, industry totals for 1880 are far from accurate. Still, the census remains the best source for understanding the men and women who had, by 1880, built a mature and profitable industry in South Carolina.

The 1880 census demonstrated that of those land miners enumerators detected, or "visible" miners, the monopoly of mature black males remained in place. Ninety-seven percent of all miners tallied were black or mulatto males and their average age was 31.6, an increase since 1870 in a state whose male population had grown younger. Bucking the male-dominant trend, four female phosphate laborers, ranging in age from seven to eighty, appeared in the 1880 census schedule. Atypical in sex and age for the work,

Figure 7. Chimney remains and decorative plantings from single-family house dating to the hand-mining era, Bulow Mine, 2006. Photo by author.

the four likely had lighter responsibilities at the washers or mines. Each had a male relative or housemate miner who likely facilitated entry into the industry and protected them in the field.[73]

The big news from 1880 was that visible miners dramatically changed their living arrangements from the previous census. While enumerators found most land miners living in group-housing arrangements in 1870, they failed to find any a decade later. In 1880, 63 percent of miners, laborers, and hands were married, and most lived with their wives and children (1.85 average) in single-family dwellings. Over half of the wives earned wages (mainly as field laborers) and over a quarter were "keeping house." Single miners either lived alone, with family, with other miners in small groups, or boarded with other families.[74] Several factors contributed to the shift in living arrangements, including migrating mine locations and the developing black community. As Ashley River deposits became depleted, mines moved farther inland, away from plantations such as Middleton Place and Drayton Hall and closer to miners' homes in rural hamlets like Red Top to the north and west. As mining grew so too did surrounding black communities.

Although land miners appeared more settled in census data, their families still did not adhere to the Victorian ideal. Two-thirds of the miners' wives worked for the family economy and farm wages and did not withdraw from the fields. As in 1870, the 1880 population survey gave a tantalizingly incomplete view of the black economy, of which phosphate mining was merely a part.[75] Other evidence from the phosphate-mining counties indicates that the task and two-day systems survived into the 1880s. In the new Berkeley County, carved out of rural parts of Charleston County in 1882, most field laborers worked by the task, and the two-day system, with access to eight to ten acres of land and a house, was "most in vogue." Colleton County's black laborers were evenly divided between wage contracts and a variant of the two-day system. Most of Beaufort County's "colored" farmers owned farms, but many, especially those on St. Helena Island, combined agricultural work with phosphate (river) mining.[76]

Non-census sources revealed more about the quality of miners' living and working conditions. Most freedpeople seemed to prefer living in newer houses and rejected slave cabins, but one newspaper account painted a bleak picture of miners and their houses. Describing a "general squalor and untidiness," a *New York Times* correspondent found rooms in miners' cabins "equally dirty and comfortless." She described the phosphate workers as "sullen and silent" at home and in the trenches and the women with "a pipe between their lips." The writer also perceived a strong distaste for the "hard, . . . unhealthy, and unpleasant" work. Explaining why he failed to work regularly, a worker told her "It too much tiring to work ebery [*sic*] day." The reporter heard many complaints about low pay, but she noted that superintendents justified the wages as "all the uncertainty of their labor deserves."[77]

Despite its absence from the census, group housing persisted into the 1880s. Enumerators canvassed the mining district in early to mid-June, so they missed the laborers who migrated to group housing during the summer lay-by. They also ignored the state's convicts leased to phosphate-mining companies. The year 1880 marked the start of the convict lease's "most flourishing years" in South Carolina, and mining played a major role in the exploitative system. State records reported in 1880 that almost 10 percent of the state's convict population (58 of 590) mined at the Cahill & Wise Phosphate Works. The following year, mining accounted for over 84 percent of the state's leased convicts. Cahill & Wise paid the state $12.50 monthly for each of its 82 convicts mining near Charleston, while R. S. Pringle paid $10 and used 120 at a Summerville mine.[78] As they did with other miners, companies housed convicts in existing cabins or new barracks. Housing records from Cahill & Wise and Pringle are not available, but records from Drayton Hall (Charles H. Drayton & Co. or CMMC) indicate that convict miners slept in former slave cabins.[79] Along with migrating workers, convict laborers were another important group lost by the census.

Although convict lessees paid very little in the early 1880s, most companies paid free miners more than they could make as farm laborers. Employers reported to the census that the average day's wage for skilled mechanics was $1.00–2.50 and for ordinary laborers $.75–1.00. Miners earned $.25–.30 per vertical foot, so "industrious" miners took home $1.00–1.75, although some took less pay with daily rations. Mining paid comparatively well. Male field laborers in the mid-1880s working in the mining counties earned a daily wage of $.50–.75, and women earned $.40–.50. Men with monthly contracts earned $8–10 with rations, and women earned $4–6. Statewide, farming wages averaged about $8–9 per month for men and $5–6 for women.[80] Mining was exhausting work, but the superior pay made it a vital part of lowcountry freedpeople's economic strategies.

Good wages did not translate into better educational prospects. Phosphate workers gained in literacy during the 1870s, but the statistics remained abysmal. Whereas less than 1 percent of phosphate workers in St. James Parish Goose Creek could write in 1870, the rate improved to 10.3

percent by 1880. Only 8 percent of miners in St. Andrews Parish could write in 1880. The men from both mining parishes narrowed the literacy gap with the state and city averages during the 1870s but still trailed substantially. South Carolina's blacks had a literacy rate of 21.5 percent in 1880, while Charleston's rate stood at 50 percent.[81]

While informative for the lives of visible workers, the 1880 census failed to accurately count the number of miners. Manufactures and population schedules varied widely in their tallies of all phosphate-related workers, and enumerators did not clearly distinguish between land mining, river mining, and fertilizer manufacturing. Companies reported that the "greatest number of hands employed at any one time during the year" totaled 3,031, while only 820 employees identified themselves as working in the three industries. Use of the term "greatest" instead of "average" inflated the manufactures schedule numbers, and failure to account for part-time workers deflated the population schedule numbers. Interpreting the manufactures data by location and company yields a total of 1,685 land miners, but similar analysis of the population schedules' occupation listings, as well as context and location, identifies only 449 miners. The higher estimate is likely the more accurate one, given the continuing ambiguities in lowcountry laborers' occupations. Like its predecessor, the 1880 census compendium ignored phosphate mining and likely included miners and firms as part of the fertilizer list. Its fertilizers category listed the average number of hands as 2,759 in 28 establishments, two numbers suspiciously high for just that industry.[82] Once again, the census rendered land mining virtually invisible.

As in 1870, the 1880 census' problems with the land-mining industry began with enumerator inconsistency. Census takers randomly listed workplace, task, and raw material for workers' jobs. Unclear terms such as "phosphate," "works," "mill," and "digger" caused confused census statisticians to group all workers in the fertilizer category. In reality, an individual listed as "phosphate mill" could have worked in a fertilizer factory or washed newly mined rock at the river. A "phosphates" listing described both miners (and associated jobs) and factory workers. Entries such as "phosphate hand" and "phosphate laborer" contained residual imprecision. Even the seemingly obvious listings "phosphate miner" and "phosphate digger" were problematic, because river mining, which had grown dramatically since 1870, and land mining took place in all three counties. Companies and managers suffered from similar census ambiguities.[83]

Unfortunately, no other reliable list of land-mining workers or companies survived from 1880. Such a list would have enabled the historian to estimate, based on production per miner, the number of workers missed by the census. Since non-census sources from contiguous years indicate that many more land companies existed (twenty-seven) than appeared in the manufacturing schedule (twelve), it is reasonable to assume that the land industry had at least 1,685 workers and perhaps as many as the *Courier*'s 1884 estimate of 3,967.[84] Despite census confusion, it ranked high among South Carolina's other industries. Based on the conservative 1,685 number, land-phosphate mining ranked second in the state's industries employing male adults, behind the tar and turpentine industry (4,512 men) but ahead of lumber (1,431), flour and grist mills (1,048), and cotton goods (661).[85]

Hand mining began as a prolonged battle against nature and matured as a series of evolving processes in which manager and miner negotiated a balance between efficiency and independence. Building on the history of task slavery, freedpeople created a relatively autonomous labor system and demanded control over the meaning of freedom in their home and work lives. Only by preserving the principle of worker flexibility and offering good wages could managers coax the reluctant ex-slaves back into the ditches. Freedmen adapted the task system to postbellum realities, working for someone else part of the week and for themselves the remainder. This permitted men to farm and mine on a weekly, monthly, or seasonal basis. Flawed and incomplete, census data from 1870 and 1880 aid the historian in partially illuminating the black labor force, in terms of age, literacy, living conditions, and wages. More fundamentally, the gaps in census data suggest the unique nature of a lowcountry economy in which phosphate played an important role in the freedpeople's transition from slavery to freedom. Finally, Williams Middleton's trials as slave owner turned mining entrepreneur demonstrate the magnitude of change for postbellum planters, the effectiveness of freedpeople's negotiations, and the challenges in harnessing this free-flowing mass of labor to a single industry.

4

River Mining and Reconstruction Politics, 1869–1874

Just as South Carolina's planters and slaves had grown rice in both dry and water cultures, so too did the state's entrepreneurs and laborers mine phosphate on land and in the rivers. River mining unofficially began in 1869 as an unsanctioned use of public resources and officially in 1870 as a separate industry from land mining. Although a few land-mine entrepreneurs entered river mining, the new industry had mainly different owners, laborers, and locations. Landownership was not a prerequisite for river mining. Most laborers came from the Sea Islands, and most river mining took place closer to Beaufort.[1] Finally, the rivers were public domain, so mining was subject to direct state regulation and taxation. Indeed, widespread debate in legislative halls, on newspaper pages, and at public meetings over royalties, mining rights, riparian rights, and navigation made regulation of this industry one of the most contentious issues of the day.

The river-mining industry emerged during the politically chaotic years of Reconstruction. With ex-Confederates disfranchised, the Republican Party had the opportunity to reshape white society and help advance freedpeople. Hardly unified, however, the party was a loose collection of suspicious factions vying for control. Carpetbaggers, scalawags, northern blacks, and freedpeople fought, sometimes bitterly, over civil rights and financial legislation. Much of the river-mining industry's early history, including the creation of its first few companies, involved maneuvers within various Republican factions and unusual alliances with some Democrats. The battles also involved confusing alliances based on class and race. Ideology and self-interest often merged when Republican legislators met, a fact not lost on Democratic businessmen. Carpetbaggers sought to improve

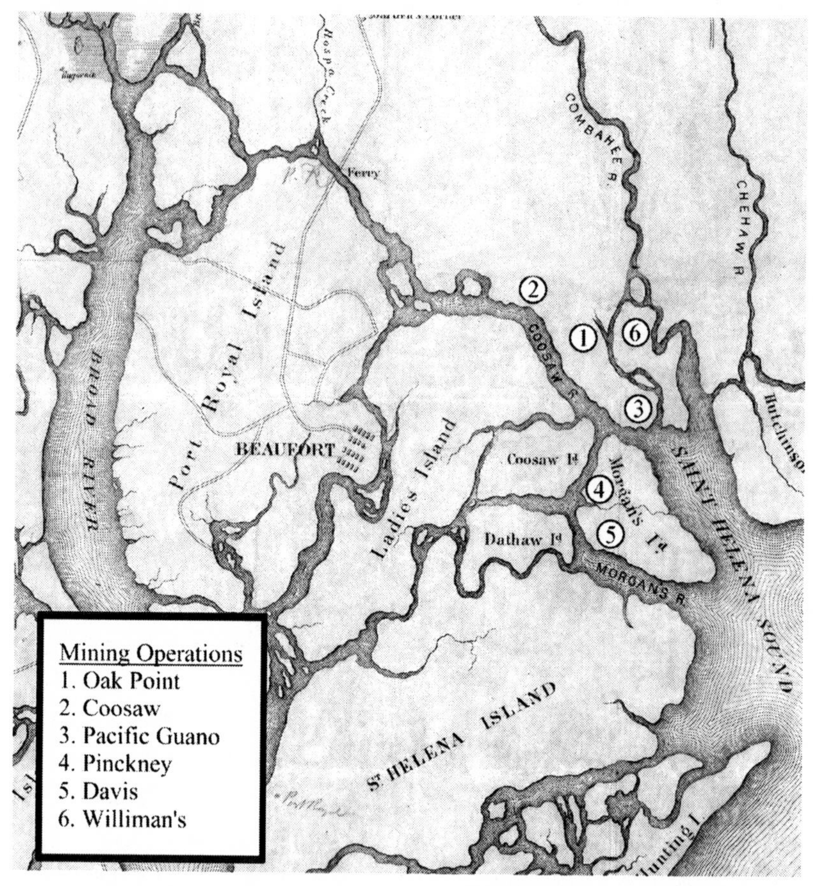

Map 3. River-mining territories. Map created by author.

commerce by working with local entrepreneurs to establish large, potentially monopolistic river-mining firms. Attempting to construct a new society, black leaders fought to stop the monopoly by forming their own companies. And Charleston's most influential newspaper used the monopoly issue to advance its conservative agenda. The contentious debates would outlive passage of the first river-mining act and reverberate throughout the next several decades. Riding out the legal and legislative turmoil, black Sea Islanders shouldered the brunt of the work and negotiated their best interests.

The Road to Regulation: 1869

An outgrowth of the land-mining bonanza, river mining began in South Carolina before it was legal. Throughout 1869 and possibly earlier, "a large number of poor men," mostly freedmen, gathered phosphate rocks from the state's waterways. Picking from the shores at low tide, using oyster tongs in shallow water, or diving in deeper waters, the men transported the rock to Charleston or Beaufort in canoes, sloops, or schooners and sold it to local fertilizer manufacturers or land-rock companies. Estimates in late 1869 had 1,200 men earning five dollars per day—superb pay for a workingman—harvesting "thousands of tons" of rock.[2] For the freedmen, the job was ideal, furnishing extra income without white supervision during the slow summer months. For manufacturers and rock dealers, the quiet trade increased their supply of phosphate without the complications of payroll, supervision, and land taxes. Unfortunately, state legislators noticed.

The origins of river mining exposed basic differences between land and river mining and raised fundamental questions for lawmakers. While phosphate lands belonged to private owners who paid property taxes, the riverbeds belonged to the people of the state. Some form of access to river rock was a public right, but its removal required compensation to the state. The rock was not uniformly distributed throughout the state or even within the same river. Tightly packed beneath lowcountry waterways, some reserves required expensive dredges to break up and hoist the rock to the surface. Other areas had scattered nodules easily harvested by hand or rudimentary tools. As for state regulations, few precedents existed. Unlike fish, river rock was neither migratory nor renewable. It was not a navigation issue. How then to regulate and tax river mining? To whom should charters be granted? The legislature's decisions shaped all aspects of the new river-mining companies—their personnel, mining territory, and other rights. Further complicating the issue, many lawmakers allowed greed to influence their votes. Bribery was rampant and conflicts of interest common.[3] Not surprisingly, then, deliberations in the statehouse and partisan newspapers exposed many of the state's political, economic, and racial fault lines. Within a flood of rhetoric flowed ideological agendas often indistinguishable from naked grabs for power and money.

Republican state Senator Stephen A. Swails ignited a firestorm in December 1869 when he introduced the first river-mining bill. He proposed to grant to associated individuals, including Charleston businessmen George W. Williams and Edward Willis, the "exclusive right" to establish a company to mine phosphate rock from the state's waters and pay a twenty-cent "royalty" for each ton removed. Republican Daniel T. Corbin soon emerged as the bill's leading advocate.[4] In the House, Republican George M. Wells sponsored a bill that included thirteen legislators, ten of whom were black Republicans.[5] These legislative conflicts of interest failed to stir the state's turbulent waters as much as the single phrase "exclusive right." Should the state grant one company the exclusive right to mine all its rivers? Did this constitute a monopoly?

The editors of Charleston's leading Democratic newspaper, the *Daily Courier*, seized on the river-mining bill as an opportunity to advance their agenda, the economic interests of their readership, and the cause of anti-Republican obstructionism. Purporting to champion the interests of capital, labor, and taxpayer, the editors targeted Republican legislators' creation of an "odious monopoly" and the inadequate royalty, arguing that the bill would destroy existing land-phosphate companies unable to compete with the "small 'ring' of selfish capitalists." They even posed as defenders of the state's "laboring men . . . especially the colored portion," fretting that the bill would deprive freedmen of jobs and food.[6] This ruse was particularly transparent, coming as it did from the elite, white-supremacist, and Democratic newspaper not known for its regard for black workingmen pursuing nonagricultural and relatively autonomous jobs. The editors' real purpose was to protect the interests of Charleston's business and planter aristocracy, many of whom were part of the land-mining and fertilizer industries and directly benefited from unregulated river-rock mining.

Behind the *Courier*'s language lay a desire for tax relief for, and control of the new industries by, elite Democrats. The editors argued that the proposed royalty was inadequate to relieve citizens' tax burdens; by "citizens" they meant plantation owners. They pushed for a fifty-cent royalty and annual mining licenses to increase state revenues and for licensing to compel minimum yearly production levels, thus decreasing planters' fertilizer costs.[7] The editors' concerns for the farmer and the state budget were touching indeed, but their primary goal was to preserve as much of the status quo as possible. If the days of Democratic elites acquiring cheap, unregulated river rock for their land-mining and fertilizer companies were over, then they and their *Courier* allies sought to have state regulators limit the new industry to only small, less powerful companies.

Seeking to mobilize popular opposition, the editors linked the bill's "exclusive rights" clause to monarchical excesses, such as kings' "absolute right" over fishing and navigation.[8] Antimonopoly rhetoric was effective for discrediting Republican rule, especially after the onset of Congressional Reconstruction, but when paired with fishing or navigation, it only confused deliberations. Since the rock was not a renewable resource, the major issues were the rate and extent of its depletion. In order to fully benefit the fiscally troubled state, depletion had to be finished before phosphate discoveries in other states or countries undermined South Carolina's dominant position. And the mining had to be thorough; a company unable to completely mine its territory would waste the state's resource. As for navigation, miners were not allowed to obstruct any waterway, and their activity only aided navigation by deepening channels.

However, the editors' denunciation of the bill's monopolistic wording was on the mark, identifying the bill's fatal flaw and the industry's main controversy. While various companies would try to defend it, the granting of one company exclusive mining rights lacked not just equity but logic. No one company could adequately explore and mine the entire state's phosphate deposits. The more relevant issues were territorial exclusivity and the number of licensees. The supply of easily reached rock was limited, and large capital was necessary for removing deep, embedded rock. Advocates of unlimited licensing with no specified territories trumpeted democratic and capitalistic ideals and argued that too few companies might leave most deposits untouched for decades. Supporters of restrictive licensing and defined territories feared that too many companies would discourage large capitalists from investing in substantial plant. It became a debate about technology and size. Was dredging the most beneficial method for state revenues, or was hand mining with simple tools just as good? Was it fair for rock loosened by dredges to be collected by other companies' miners? Would an industry composed of lightly equipped companies ignoring deep deposits cost the state substantial revenues? Would enforcement be easier with fewer companies?

The *Courier's* warnings motivated Democrats as well as Republicans, and other newspapers throughout the state adopted the crusade. In December 1869, the *Courier* reported that four hundred citizens, black and white, rallied in Charleston against the growing phosphate "ring." Protesters argued that the bill would make an already bad situation—what they described as "grievously low wages" due to an overabundant supply of labor—worse by eliminating another category of independent and decent-paying work. In addition, "small capitalists" would be "at the mercy of the monopolists." By ending competition and choosing favorites, the proposed bill would violate "the spirit of a republican Government." Foreshadowing his party's split, Republican Judge Thomas Jefferson Mackey argued that the bill was an article of "public plunder" and "against the interest of the working people." Hardly a friend of the *Courier* and the local elite, he nevertheless branded Corbin a "monstrous bag of wind and wickedness" operating within a bribed legislature.[9]

The "monstrous bag" urged legislators to end the illegal removal of river rock, "encourage" the development of a regulated industry, and offer "proper inducements" to large investors. Much of the rock, Corbin believed, lay in deep water where only machinery "at large expense" could reach it, so designing an industry with high royalties and low bonds for hand miners was senseless. A modest forty-cent royalty would facilitate the creation of substantial companies capable of deep river mining, despite its greater initial expenses. The sizable $25,000 bond would ensure "true and faithful" returns to the state. Cultivating large and honest mining companies, he reasoned, was the state's best route to maximizing its royalty revenues. Shutting out small companies, including those run by freedmen, was a small price to pay for fiscal responsibility. Responding to cries of monopoly, Corbin argued that those named in the bill were Charlestonians with experience mining phosphate and that they came to the legislature with an "honest and square business proposition."[10]

Corbin assumed the loyalty of freedmen voters, and his faction's businesslike arguments disturbed black Republican legislators. Ex-slave and native W. Beverly Nash was one of several black senators attempting to block Corbin's plans. He was not eager to bestow exclusive rights on anybody, especially businessmen like Willis who were some of "the strongest opponents to the workingmen's movement." A procedural vote on the phosphate bill illuminated the contours of this growing inner-party rift. Within the "pro-bill" group, eight of the nine were whites (of whom half were northerners) and all were Republicans. Only four of the nine "anti-bill" senators were white, none were northerners, and two were Democrats. Meanwhile, Corbin tried to repair his reputation with labor and its advocates by arguing that public outcry about the rights of the small entrepreneur or individual miner was misplaced. Only large capitalists with steam dredges could replace scattered rock thieves with "five or six hundred men getting honest wages." Proper incentives and guarantees would convince "responsible parties" to invest capital, provide regular employment, and provide the state income. Nash remained skeptical about Corbin's "responsible parties," and the trickle-down scheme left him unmoved.[11]

Corbin then took aim at the bill's other enemies. Citing the "infernal howl" raised by the Charleston papers, he traced the source of the opposition to local land-mining entrepreneur George S. Cameron and other "gentlemen" who had benefited from the regulatory vacuum and illegally mined ten thousand tons from the Coosaw River. They "never whispered a word," Corbin claimed, until an "honest" proposal came forward, and only then did they "call upon the people to rise against all monopolies." Corbin's attack exposed yet another element of the river-mining debate, that of naked rivalry among white, northern, business-minded competitors. Cameron's backer was Simon Cameron, the Pennsylvania senator whose Republican faction was moving the party closer to big business and away from social reform. Despite Cameron's alleged hypocrisy, Corbin defended big business by observing (perhaps more candidly than he intended) that protests against the bill represented "on the one side . . . the cry of the poor man, and on the other the rich man. It is nothing but the old cry—capital and labor." According to Corbin, well-capitalized firms with experienced managers not only provided the most steady work for laborers but also did the best job for the state.[12]

Corbin concluded his defense by citing the state's decades-old agreement with the South Carolina Rail Road Company (SCRR). The *Courier*'s editors remained unconvinced. They gave no credence to Corbin's attempt to fashion the SCRR's tax-free perpetual charter into a precedent for exclusive rights in river-phosphate mining. Chartered in 1827, the SCRR acquired state backing in the 1850s. *Courier* editorials correctly noted that the legislature had not given the railroad exclusive rights to run all potential railroad routes in the state as the phosphate bill did with mining the state's rivers.[13] More importantly, the Corbin-*Courier* debate incited more opposition to the bill by exhuming bitterness over earlier battles within the state and the South over common rights and private control of public areas.

Although not strictly an issue of riparian rights, the phosphate-bill controversy was reminiscent of antebellum struggles involving landowners along rivers, early industrialists, and increasingly business-friendly legislators. These struggles established the context within which Corbin, the *Courier's* editors, and others waged what would become a war over the rights and inducements given to the river-mining companies. South Carolina's leaders initially supported notions of traditional political economy. In acts passed in 1823, 1825, and 1829, the General Assembly created a balance of public and private interests by ensuring that the state's waterways remained open to all. State courts agreed, ruling in 1822 that no one entity could have a monopoly on a river, even on those not yet navigable. Tom Downey reveals that this consensus deteriorated over the next few decades, and a fundamental rift developed between "men of property" and "men of capital." The former were sawmill operators and other businessmen who maintained close ties to the state's traditional agricultural economy through ownership of land and slaves. Men of capital included corporate investors who organized enterprises such as William Gregg's Graniteville Manufacturing Company and whose wealth was more liquid. An 1853 act broke precedent by granting to Gregg's factory what amounted to exclusive rights to control the water flow on a section of Horse Creek. Legislators had become more willing to cast aside tradition to assist men of capital in promoting large-scale industrialization. "The legislature," Downey argues, "no longer merely regulated competition . . . ; instead it chose the winners and losers."[14]

The 1853 act set three important precedents bearing on the 1869 river-mining debate. First, legislators granted to a corporation exclusive rights to a river section. This harmed other business owners and free competition, but legislators believed it was necessary for the greater good, bringing large industry to the state. Second, the inducements broke well-established legal precedents by allowing owners of a company to regulate navigation to benefit their specific business. Third, the 1853 act contained an element of social control by promoting industrialization and the creation of nonagricultural occupations for the state's landless white population. The phosphate bill was more moderate than the 1853 act, because it offered smaller inducements to still-unformed companies in an industry of still-indeterminate size and value and did not interfere with navigation or the legal activities of established companies. The 1869 bill also rested firmly on the 1853 precedent of exclusive rights (if interpreted as within a specific territory), and it promoted social control for the large, landless population (now black) in need of organized industrial employment.[15]

Controversy over southern stock laws provided another layer of context for the river-mining bill. Steven Hahn describes antebellum hunting, fishing, and open-range foraging as important "common rights" that planters were unsuccessful in regulating. After the war, planters fearing outbreaks of mobocracy renewed their push for stock laws, demanding that all farmers control their livestock and ending crucial common rights for freedpeople. The new laws formed the basis of a redefinition of property as "absolute and exclusive." Black and white opponents viewed stock laws as "instrument[s] of class oppression" and evidence of elite efforts to make small producers dependent. For opponents, stock laws represented the supreme test of democratic rights and state power and a showdown between productive property and accumulated wealth.[16] The language of the stock-law debate permeated arguments about "exclusive rights" in the river-mining industry. "Producers" complained that the lowcountry aristocracy sought to obtain special privileges and make the poor dependent. Planters, lawyers, and factors entering the river-mining industry feared unregulated mobs stealing rock from the rivers. Building on long-simmering resentments, upstate politicians attacked Charlestonians and the lowcountry elite. All sides questioned the influence of the opposing classes in government and debated the relationship of exclusive rights to monopoly.

South Carolina's antimonopolists of 1869 drew on, and continued, a rich tradition. Opposition to state-granted monopolies and desires to restrain the state had been well-established parts of Anglo-American culture since the reign of Elizabeth I. Throughout nineteenth-century America, antimonopolism evolved into a popular ideology for coping with political and economic changes, and, in the late 1860s, three related tenets emerged as the foundations for that ideology. The first involved the challenge of "producers"—including small businessmen, laborers, and farmers—to resist the schemes of wealthy nonproducers to make them economically and politically dependent. The second and third tenets included fears of corruption growing out of the postwar banking and currency systems. After the war, antimonopoly sentiment became focused on corporations, their owners, and the corrupt "rings" they formed with politicians. Antimonopolists accused corporations of "stifling opportunity" for small entrepreneurs, and they distrusted defenders of industrial capitalism, including Corbin, who championed "natural" monopolies.[17]

The year 1869 was an especially important one for antimonopoly fervor. *Courier* editors initiated their crusade the same year that elites Charles and Henry Adams began attacking monopoly, tyranny, and industrial capitalism in New York. The sense of noblesse oblige that informed the Adams' reform was firmly entrenched in Charleston's Democratic business circles, where some elites denigrated industrialization as not being a "proper" profession and its grasping entrepreneurs as less than honorable. The year featured national headlines about railroads, railroad barons, and attempts to regulate them. Jay Gould, Jim Fisk, and Cornelius Vanderbilt became symbols of decadence, privilege, and dishonest wealth; Gould's 1869 attempt to corner the gold market added to popular antipathy. Railroad reformers spoke a language of antimonopoly that South Carolinians with various motives embraced. River-mining bill opponents, including some elites, lumped Corbin and the pro-bill entrepreneurs together with Vanderbilt, Gould, and Fisk. Rumors of bribery and rings surfaced in the Palmetto State, just as they did in the Empire State. And the birth of railroad-reform movements in Massachusetts and Minnesota encouraged the bill's opponents in their quest to stop government-granted monopolies.[18]

The Reality of Regulation: 1870

The *Courier* and its readers continued to reflect and shape the debate in 1870. "Free Trade" wrote to chastise Corbin for directing his "sneers and satire" toward "a few poor fisherman" whose access to river phosphates was "inherent, and to be enjoyed in common." The writer noted that the carpetbagger Corbin avowed "liberty and equality for your new friend, the colored man," but, in advocating monopoly, denied him the fruit of "his toil and the labor of his own arm." The *Courier*'s editors then introduced new arguments involving "common utility," asserting that a river-mining monopoly would be "disastrous" to the state's best interests.[19] Unfortunately, editors, writers, legislators, and political operatives failed to clarify just what the common utility was. As state leaders gave birth to river-mining regulation, questions of ideology, sectionalism, party politics, race, class, and corruption further muddied the waters, resulting in a deeply flawed act.

The 1870 legislative session revealed political divisions and practical questions among "exclusives and anti-exclusives" within both parties. Maneuvering between carpetbaggers and African Americans foreshadowed increasing conflict with the Republican Party. The other party appeared equally divided. While "moderate" Democrats had begun to make political overtures to "reform" Republicans, merely to split the latter's party, some Charleston Democrats made real alliances with Republicans to facilitate business deals. Rumors of influence and corruption swirled in Columbia and Charleston. Senator Charles P. Leslie suspected that behind the river-mining bill lay a Charleston "ring" led by Corbin, and he lamented the lack of available information about the rock's value, the deposits' size, or a company's costs. Leslie worried that an excessive royalty would promote underreporting, while a low one could cost the state revenue and not necessarily lower fertilizer costs. As for proposed licensing systems, he argued that a "highest bidder" scheme would prolong the "exclusive" debate and produce "just as much rascality." A "general license" plan would spur competition, but regulation would be more difficult with multiple companies. Senator Henry E. Hayne demanded the bill's postponement until scientist Charles U. Shepard Jr. reported the deposits' quality and size, but Leslie dismissed Shepard as a "humbug" who the Charleston ring would control. Hayne also touched upon what would become a great controversy within the industry, the issue of whether marsh mining took place on land or in the river.[20] The marshes filtered river water but, like rice fields, were real estate, so there was no regulatory consensus.

Meanwhile, black antimonopolists increasingly employed the language of class in promoting their constituents' interests. Republican Richard H. Cain contrasted the rights of the wealthy and their corporations with those of common (freed) men: "There is a time when the rights of the poor man should be recognized; when we should legislate for their interests, and not undertake to give exclusive rights, as this Bill proposes, to rich men." To do so, he argued, would "take bread out of poor men's mouths." Although reasonably opposing the concentration of power inherent in monopoly and exclusive rights, Cain's words troubled a growing number of upper- and middle-class whites who believed working-class advocacy to be dangerous. Hailing from both parties, the "better classes" saw themselves as the only citizens dedicated to individualism and hard work. Rhetoric from white Republican political insiders, such as Timothy Hurley, who described state politics as "Pine Knots versus Huguenots; Shovelry versus Chivalry," was equally disturbing. The alleged class war dated back to the 1868 black-majority legislature that passed tax- and land-reform legislation. By 1869, panicking whites believed that the legislature had been captured not only by blacks but also the lower classes intent on instituting pro-labor and anticapital laws. Striking Charleston longshoremen and the black state militia confirmed conservatives' suspicions of an uprising of society's lower ranks.[21]

Despite the fact that most of the state's black legislators vigorously opposed the river-mining bill and that many carpetbaggers supported it, white observers throughout the nation erroneously considered the "exclusive rights" debate further proof of class warfare. Widely publicized by South Carolina's Democrats and later adopted by northern newspapers, these fears grew into a national consensus and contributed to what Heather Cox Richardson terms the "Death of Reconstruction." The better classes were especially outraged at the prospect of the lower orders hijacking state government to establish a monopoly. They rejected monopolies as evil concentrations that violated free labor theory by creating a permanent class of wage laborers denied upward mobility. Southern conservatives opposed monopolies just as fervently because they threatened white male independence. For the northern and southern better classes, South Carolina's chaotic political scene seemed dominated by corrupt Republicans, including demagogues who, aided by a black majority, were rallying workers with calls of class conflict and wealth appropriation.[22]

But South Carolina Republicans were not as unified as the better classes feared. By mid-January 1870, the river-mining bill debates exposed the Cain-Corbin split to be indicative of an emerging "new programme to create a party of color" among black Republicans. Distrusting carpetbaggers and scalawags, black leaders aggressively demanded statewide political office. With the South's largest black majority, South Carolina's blacks were unique in their political assertiveness. Thomas Holt demonstrates that black and white Republicans voted differently in 38 percent of critical votes during 1869–70. Even within their own ranks, black legislators were not a unified voting bloc, except in civil rights matters. They were divided on financial matters such as mining regulation, and their color, wealth, and social origins mattered. Wealthy mulatto freemen were more likely to vote conservatively, while black lawmakers of slave or free origins were more radical; mulatto former slaves were less predictable. Corruption did not alter these patterns, but it did lead to a significant Republican split in the 1870 election, one that added another layer of context to the phosphate-bill debate. Seeking to regain the moral high ground, "Reform" Republicans broke off from the "regulars" to form alliances with opportunistic Democrats. Cain and others "bolted" to the Union Reform Party in the spring.[23]

Black Republicans and Corbin continued their tactical battles in early 1870. Seeking to undermine Corbin's "exclusive" bill, Cain introduced a Senate bill to incorporate and license the South Carolina Chemical and Mining Company (S.C. Chemical). Corbin responded by fostering fears of a conspiracy by Nathaniel A. Pratt of Charleston Mining and Manufacturing Company (CMMC) to sabotage the river industry by altering the original bill—now "Dr. Pratt's Bill"—by increasing the royalty to prohibitive levels. No friend of Pratt, Cain raised the stakes by recommending a two dollar royalty, significantly higher than previous proposals, and territorial restrictions, the first such stipulation. Senators defeated both proposals. Corbin and supporters finally dropped the "exclusive rights" phrase from the bill, believing that the new entity would remain the state's only mining company and therefore still "exclusive." Opponents disagreed, celebrated their victory, and moved on to other fundamental details. Nash distrusted Willis and Alexander R. Chisolm on labor issues and removed their names from the bill. Hayne and W. R. Hoyt attempted to shorten the bill's term, but the Senate agreed to Corbin's motion for a twenty-one-year term to encourage substantial investment. Senators grappled with the royalty, eventually settling on one dollar per ton. In February 1870, both houses passed what the *Courier* persisted in labeling "the exclusive right bill."[24] Not surprisingly, the words "exclusive," "term," "territory," and "royalty" framed all future debates concerning the river-mining industry.

Republican Governor Robert K. Scott vetoed the bill he described as "a naked grant . . . to a few individuals." He worried that the bill contained no production minimums, sanctioned a monopoly, and raised farmers' fertilizer costs. Citing the alleged "Pratt Bill" conspiracy, he warned about potential sabotage of the nascent river industry. In short, the notoriously corrupt Scott believed that the bill was the product of bribes and benefited powerful rich men over poor workers. Indeed, *Courier* reporters noted that Corbin's lieutenants carrying carpetbags of greenbacks enabled the bill to pass "with that peculiar glide which showed the power of 'grease.'" But legislators overrode Scott's veto by a wide margin.[25] Lobbyists, businessmen, and their money played an important role in the phosphate law's passage, and in the politically charged atmosphere of Reconstruction, corruption became a central part of its legacy.

Courier reporters offered tantalizing silhouettes of cash-laden lobbyists, but did not identify them. Witnesses later identified "the leader of corruption" to be Timothy Hurley, a carpetbagger, state legislator, and Republican newspaper owner living in Charleston. Hurley allegedly telegraphed George W. Williams, picked up sixty thousand dollars from the Scott, Williams & Company Bank, and carried a "large bundle" for distribution to legislators. Willis later admitted that he too was "instrumental" in securing passage. Williams and Willis were part of Charleston's traditionally Democratic business community, who reached agreements with Republican legislators for personal gain and ideological reasons. They believed that the new industry had great potential and that the state's elite should control it.[26]

The Industry Evolves

In 1870, the Marine and River Phosphate Mining and Manufacturing Company of South Carolina (Marine) began operations as the only company licensed to mine the state's rivers. It became even more exclusive when the S.C. Chemical bill died on the Senate floor. Worse yet for black legislators fighting for the working class, Marine soon became a creature of Williams' Charleston empire. Banker, cotton factor, commission merchant, and diversified businessman, Williams knew that more phosphate rock on the market meant cheaper fertilizer (for his Carolina Fertilizer Company), more cotton (for his factorage business), and more financial activity (for his banks). He also was an elite Charleston booster who undoubtedly preferred that society's traditional leaders remain in control. Most of Marine's other founders were southerners associated with Williams' businesses, but the two most significant ones were northerners.[27] William L. Bradley was Boston's best-known superphosphate manufacturer, a small land-rock producer near Charleston, and a leader among Marine stockholders. Charles C. Coe came from a northern family of superphosphate manufacturers and played a vital role as Marine's first superintendent. Williams likely met Coe and Bradley when he began dealing fertilizer in 1869.[28]

Coe was Marine's point man. While land-mining superintendents primarily managed labor, their river-mining brethren managed all stages of the production process. Coe oversaw laborers, the physical plant (buildings, boats, and machinery), and rock preparation. He also weighed the rock, calculating totals for sales and royalties. Coe chose when to purchase additional plant and where to ship the rock, and he negotiated with purchasers. Finally, he was the one responsible for protecting the company's interests during dredging, including chasing off, and, if necessary, filing suits against, trespassers. Backed by Marine's affluent stockholders, Coe spent two hundred thousand dollars on steam-powered dredges, washers, crushers, tugboats, lighters and flats, and he had the company mining by May 1870. Due to census deficiencies and a paucity of sources, little is known about early river miners. Coe likely hired men who mined illegally before 1870, and experience on the lowcountry's intricate river systems became especially valuable as the search for phosphate expanded beyond the Ashley River. Marine's miners dredged the Stono River near Charleston and the Bull River and North Wimbee Creek north of Beaufort, and Coe bought or built wharves, storage buildings, and grinding mills at the latter location, as well as near the Beaufort River. Marine mined 1,989 tons for the year ending in October 1870.[29]

In spring 1871, scientist Charles U. Shepard Sr. joined Coe on Marine's steam tug and furnished a rare account of the dredging and washing processes. He observed of the mined rock, "On being let fall upon the deck of the washing tug, contiguous to the hopper of the washer, it was worked over with hoes and shovels, and the coarser lumps, together with a portion of the finer gravel, introduced into the washer; while the marly and clay looking sand was pushed overboard into the river. The agitation of the

Figure 8. Dredge, crusher, and washer, Central Mining, Coosaw River, circa 1900. By permission of South Caroliniana Library, University of South Carolina, Columbia.

contents of the washer quickly separated the fine sandy material from its adhesion to the nodules, even to the bottom of their deepest holes and cavities; while the water floated it off to its former bed in the river." Typical of phosphate-mining literature, Shepard's language ignored the laborers but illuminated the efficiencies. Unlike land miners, river-rock miners could excavate, wash, and transport the rock, as well as dispose of waste, all in a continuous process. Only the grinding and drying of river rock shared similarities to land mining.[30]

Even as Marine began mining, opponents continued to attack its exclusive status. Ten independent miners, including at least two whites, tested the state's resolve or ability to enforce the phosphate act. Attorney General (and Marine investor) Daniel H. Chamberlain and Senator and District Attorney (and Marine director) Corbin sued the ten "Phosphate Grabbers" on behalf of the state for illegally mining two hundred tons on the Stono River in June 1870. Following a well-worn path, the defense argued that the "rights . . . in the 'public Domain' are inalienable," but the law and its "exclusive" intent emerged from the state Supreme Court unscathed. Marine's next challenge came from black Radical politicians, who, having lost out to the Corbin and Williams cabal in 1870, strove one year later to empower their political faction, share the phosphate windfall with their constituents, and garner riches for themselves. Legislators introduced four new bills to open the industry to small entrepreneurs by lowering license fees and bonds, and, in one case, tripling the royalty for corporations only. Two former slaves introduced what became the surviving bill. In the House, Adam P. Ford proposed to incorporate a company composed of nine men, all but one "colored." In the Senate, Robert Smalls introduced a kindred bill to create the South Carolina Phosphate and Phosphatic River Mining Company (SCPP). When passed in March 1871, the SCPP act included eight senators and eight House members.[31]

Effectively destroying Marine's claim to exclusive rights, SCPP was, in many ways, the antithesis of Marine. Legislators lengthened SCPP's term to thirty years, hoping to attract more investment. Unfortunately, SCPP's investors and founders were not wealthy or experienced in fertilizer production, so the company suffered from capital deficiencies and poor management. However, SCPP was an important step for black Republicans joining the ranks of South Carolina's commercial society. Nine of the men who formed SCPP started the Enterprise Railroad Company in Charleston, and other such ventures followed. SCPP also represented progress for black Republicans on the political learning curve. Of those listed and known in the SCPP act, all were Republicans, 75 percent were black or mulatto, 80 percent had voted for the Marine bill, and several had actively backed the SCPP bill. These statistics and the inner-party animosity suggest that after the Corbin faction outmaneuvered pro-Marine black legislators in 1870, they regrouped under the leadership of William J. Whipper, Robert B. Elliott, Alonzo J. Ransier, and W. Beverly Nash to assert their interests. Black voting on the SCPP bill was unusually united, because it was a race-based vehicle for economic advancement and encompassed the broad spectrum of black Republicans.[32]

Despite being a company dominated by black investors, SCPP had whites in a few key positions. Willis resigned from Marine's board to become SCPP's superintendent, and Republicans William Gurney and A. Canale served as president and treasurer. The leadership of a cotton factor and two commission merchants was not enough, however. Chronically underfunded, SCPP could not purchase or build machinery, boats, dredges, washers, or buildings. Instead, the company merely issued "permits" to other mining groups. SCPP's subcontractors picked phosphate rock by hand or with oyster tongs and gathered rock dislodged by Marine's dredges. SCPP then bought the scavenged rock and sold it to exporter Wyllie Campbell & Company.[33] SCPP was a marginal producer, but to Marine's investors, its very existence remained an unacceptable threat.

As SCPP and Marine warily shared the rivers, supporters began the state's first phosphate war, exposing fundamental dilemmas that would plague the industry's companies and regulators. In 1872, black legislators Smalls and Nathaniel B. Myers investigated Marine's "trifling" royalty payments, which led to the appointment of an inspector of phosphates to ensure accurate reporting. Forced to rely on voluntary reporting, inspector Otto A. Moses complained that Corbin ordered "hostile" Marine officers to deny him access to their books. Stockholder William Bradley fired back at SCPP in the courts. With large investments in Marine, he felt cheated by the loss of its monopoly, and he was alarmed by personnel changes at SCPP. By 1873, powerful white businessmen, lawyers, and politicians had taken over the once black-dominated company. Bradley also felt abandoned by Marine's leadership. Corbin had failed to stop SCPP in the Senate, and Marine's board, while sympathetic, did nothing to counter the SCPP threat. Bradley's legal team argued that the 1870 act was an exclusive-rights contract, which the state had broken by chartering SCPP. They also emphasized that since Marine's extensive operations could exhaust the "limited" phosphate supply in less than its twenty-one-year term, SCPP harmed Marine's interests. SCPP's Andrew G. Magrath responded that the 1870 act merely granted Marine the first mining license, and he cited current mining laws and court precedents in arguing that the phrase "mined and removed" in the 1870 act did not protect dredges from scavengers.[34]

Despite Magrath's arguments, current legal doctrines ignored the unique dilemmas of mining river rock, and legislators had little guidance in creating regulations. Whereas coal companies, for example, had mineral rights underneath specific acreage, Marine and SCPP had rights to the same territory. Some deposits lay in loose nodules on the river bottom, so the rock belonged to whichever company "mined and removed" it; dredges and hand picking were equally effective. However, sizable deposits existed in strata just below the river bottom, which only dredges could break. Here, Bradley's case examined the merits of mechanical versus hand mining.[35]

The key and recurring questions within the industry were: was stratum rock broken up by dredges subject to the "mined and removed" rule and therefore up for grabs, and if not, which river-bottom nodules were which? Bradley's lawyers also sought to prove that Marine's thorough and large-scale dredging (with hand mining), not SCPP's army of rock skimmers, benefited the state more. But the case was not obvious. The state's short-term interests, including its dire financial situation, dictated that unlimited miners removed the most easily accessed rock quickly. South Carolina's phosphates had gained footholds in the North and Europe in 1870, when the Franco-Prussian War limited trade, but the state's new market position was not permanent. In the longer term, however, sloppy mining meant wasted royalties, and it was unclear if adding more miners—and therefore discouraging dredging—would speed the rock harvest.[36]

Bradley's legal team argued that Marine's dredging entitled it to all nearby rock. R. R. Osgood, Marine's new superintendent, testified that dredging left debris ridges, which SCPP's "handpickers" plundered before Marine's workers could retrieve the rock. He assured the court that dredges could operate on every navigable river and that Marine combined "careful dredging" and hand picking to remove "substantially" all the rock. Bradley's lawyers concluded that dredges were a necessity for river mining. Echoing antimonopoly language, SCPP's Magrath countered that dredges were merely alternatives for large companies. Providing expert testimony for Bradley, Charles U. Shepard Jr. estimated the state's phosphate deposits to be five hundred thousand tons, a total that Marine alone could exhaust in about twenty-two years. SCPP responded with expert Moses, whose estimate was nine million tons, more than enough for several mining companies. He also described hand mining as economically feasible and dredging as "not indispensable." "Phosphate Grabbers" Henry Prince and Patrick Power bolstered Moses' opinions, explaining that they could mine ten feet deep in the Stono using heavy oyster tongs and steel borers on an outfitted boat.[37]

Several fundamental themes emerged from the Bradley case that would haunt future debates. While explicitly arguing for small deposits and exclusivity, Bradley's lawyers implicitly sought to prove that bigger was better. Dredges and well-equipped companies thoroughly and quickly mined the rivers, employed large numbers of men, and wasted little rock or royalties.

Figure 9. Wharves, Coosaw Mining, Chisolm's Island, 1880. *First Annual Report of the Commissioner of Agriculture, 1880*. (Washington, D.C.: Government Printing Office, 1881).

One big company made royalty verification and collection easier; the state's alternative was to chase dozens of subcontractors around the lowcountry's mazelike river system. Without protection for dredge operators, companies would revert to skimming easily accessible rock, leaving the strata undisturbed and royalties lost forever. Finally, Marine's large investments demonstrated the company's long-term commitment to mining. Besides arguing for enormous rock deposits and against Marine's exclusive claim, SCPP's defense implicitly sought to portray itself as the viable egalitarian alternative to Marine's monopoly.

Marine complicated Bradley's case by granting mining rights to two new companies while denying the state's right to do so. Coosaw Mining Company (Coosaw) and Oak Point Mines (Oak Point) paid royalties through Marine as subcontracting companies. Of "doubtful" legal validity, this arrangement artificially inflated Marine's royalty, which Bradley's team unsuccessfully tried to use to demonstrate Marine's value to the state. Bradley lost the case, and Marine failed to turn a profit. However, Coosaw and Oak Point became the industry's dominant companies while assuming Marine's alleged exclusive rights on two specific territories.[38]

Initially led by the Brown family of Maryland and the Adgers of Charleston, the investors forming Coosaw gained exclusive mining access in March 1870 to what would become the state's most productive river-mining territory, the Coosaw River between Lady's Island and the Bull River. They soon created a "Joint Stock Company or Mining Partnership" and leased George S. Brown's Coosaw plantation on Chisolm's Island, six miles north of Beaufort and home to a deepwater port and processing center. Coosaw's first superintendent, David Lopez, hired Charles U. Shepard Jr. to test the quality and quantity of rock, and dredging and hand picking began in late 1870. Lopez's son and assistant, Moses E. Lopez, would eventually succeed him and lead the operations for many years. Coosaw paid royalties through Marine on 1,220 tons in the 1871 production year and 5,140 in 1872. Inspector Moses gushed about Coosaw's "immense expenditures" on "extensive works" and its "systematically conducted" dredging. "Every precaution" was taken in washing and drying the rock "to establish its reputation and its value," and the company's facilities for weighing and shipment were "unrivaled."[39]

Despite its royalty link to Marine, Coosaw was an independent company whose agent, James Adger & Company, provided "very heavy advances" to the initially struggling company. The amount invested in plant (steam tugboats, dredges, washing boats, wharves, sheds, railroad cars and engines, workshops, and furnaces) and operating expenses grew to $448,000 during Coosaw's formative years. While Brown remained involved, Robert Adger became the dominant force within Coosaw, and Coosaw became a major part of the Adger family business. The Brown and Adger families had been associated since the 1790s, and by the late antebellum period, Alexander Brown and Sons and James Adger & Company were well established commercial firms that cooperated with each other and did business on both sides of the Atlantic. After the war, Robert Adger and George S. Brown maintained their families' close business relationship. The Adger family's shipping experience was an asset for Coosaw's transportation needs, while the Brown connection furnished Coosaw ties to northeastern and European fertilizer markets and manufacturers. Adger and the rest of Coosaw's board severed the link to Marine after the 1872 production year, trusting that Coosaw's exclusive rights—due to the "well established principle" of "occupation and possession"—would continue in their territory.[40]

Marine's other subcontractor, Oak Point, began in late 1869 as a land-mining company owned by South Carolinian David U. Jennings and New Yorker George W. Scott. Mining on Kean's Neck between North and

South Wimbee Creeks, just north of Coosaw's territory, Oak Point became the state's third-largest land-rock producer by 1871. Despite its agreement with Marine, the company did not pay royalties during its first several years. Scott and Jennings sold out to Wyllie Campbell & Company of London, who formed in May 1872 the South Carolina Phosphate Company, Limited, a joint stock company still known as Oak Point Mines, with a paid-in capital of £48,550. Managed by David Roberts, Oak Point helped establish Wyllie Campbell in Charleston as a domestic dealer and foreign exporter of phosphate rock. Wyllie Campbell was also an early example of what Mira Wilkins terms the "free-standing company." A form of "direct investment" and a significant international investment trend, free-standing companies funneled British and European capital overseas, registered in England but managed businesses abroad, and often provided promoters, mining engineers, traders, and industry networkers. They mirrored American river-phosphate mining companies in management structure, usually "little more than a brass nameplate" and a few directors.[41] For Charlestonians, the free-standing company represented the continuation of antebellum cotton- and rice-trading relationships with businessmen in Liverpool, London, and beyond.

In 1874, the revenue-starved state of South Carolina sued Oak Point for unpaid royalties and added it to its list of competitors. The company's troubles stemmed from the word "navigable" in the 1870 act. While Marine and Coosaw mined the always-navigable Stono and Coosaw Rivers, Oak Point mined legally ambiguous, seminavigable creeks within tidal marshes. Riparian owner Oak Point lawfully mined royalty-free rock above the low tidemark, but the state claimed that the company crossed the line, thus engaging in river mining. Again mingling private interests with public duties, Corbin won a judgment against Oak Point and collected back royalties for the state. However, the court admitted that defining the low tidemark was ultimately "impracticable" and, compromising, ordered Oak Point to pay royalties on only half the marsh rock removed in the future. This would not be the regulatory state's last success in regulating mining in marsh creeks. Meanwhile, the Republican legislature continued to regulate the river-mining industry by spurring competition and encouraging black businesses. Legislators chartered Boatmen's Phosphate River Mining Company with the same term and royalty as SCPP but with a dramatically lower license fee and bond. Similar to SCPP, it was a "company of boatmen" that tonged in any location with the company's permits. Boatmen's had no boats, machinery, or workforce, but it did undermine Marine's monopoly, empower black entrepreneurs, and employ independent African Americans.[42]

Mining and Miners

Miners and their work formed the backbone of the river-mining industry, despite their absence from government and company reports. Dredge operators shared rivers with hand miners, as well as crews for washers, tugs, lighters, and flats. Some workers manned the grinding, drying, storage, and shipping facilities on shore and in town. Others supplied firewood for the steam engines. River mining was an increasingly important source of non-farm income for Sea Islanders near Beaufort. Skilled navigators in the lowcountry's waterways, freedmen balanced mining with farming and other activities, relying on family to tend the crops. While many worked for Coosaw, Marine, or Oak Point for up to one dollar per day, others were independent contractors for SCPP or Boatmen's and were paid by the ton. Dredging continued all year, but most "hand mining," including tonging and picking, took place from April through September. Divers spent the summers harvesting loose nodules or breaking off fragments and carrying armloads or basketfuls to the surface. Flexible working conditions allowed subcontractors and day laborers to set their schedules and work pace.[43]

According to Clyde Vernon Kiser, river mining helped put largely black St. Helena Island and surrounding areas on "a sound economic basis" after emancipation. Employing about one thousand at its peak, the industry furnished a vital source of cash for local young men. St. Helena educator Laura M. Towne complained that it set "all the boys wild," including one of her former students, Walter, who left school to be a mining camp cook for five dollars per month. He returned just over a week later, "sick, crest-fallen and disappointed." Another of Towne's acquaintances, Billy Brown, suffered serious injury while living at a camp. Towne's anecdotal testimony suggests that river miners were younger than land miners and that a significant portion lived at rough mining camps instead of commuting from home. Despite her disapproval of the industry's impact on St. Helena's young men, Towne admitted that with "All the men nearly . . . at the rock . . . some money is coming back to the island." Novelist Francis Griswold offered additional evidence on the industry's impact on St. Helena: "every able-bodied man . . . was diving for the stuff. The work was cruel, but it meant fat pay envelopes for gambling and drink at Frogmore Store. The . . . women whose men worked at the Coosaw mines were beginning to go in for pin-backs, bustles, and button shoes."[44]

Life on the river was less than ideal. Supervised closely on the vessels, boat crews had less autonomy and bargaining power than their colleagues did on shore. Dissatisfaction over wages led miners to "minor unrest" in 1872. Crew members of the bark *Henrietta* staged a brief mutiny on the Bull River, refusing to go to work and jumping overboard when confronted by the captain. It is unclear if they were successful, but no other reports of labor problems appeared before 1875. Working on the rivers was also dangerous. In 1873, twelve black woodcutters for Coosaw drowned when their small boat capsized in the Beaufort River. Supplying the fuel for the company's steam-driven boats or processing machinery, the men were probably independent contractors paid by cord or ton. Untold others supplied Coosaw and its miners with tools, food, alcohol, and additional supplies.[45]

Coosaw operated four tugboats, each of which was accompanied by a dredge, a wash boat, lighters, and two small flats. Five-man dredge crews deposited rock onto the floating washer, whose fourteen-man crew then moved the cleaned rock to lighters. Crew members ate and slept on their vessels, and dredge and washer engineers had family aboard. Each flat held two "hand pickers," some of whom were Coosaw employees, but many were independents earning by the ton; one man rowed or poled while the other waded, collecting rock missed by the dredge. Most Coosaw employees worked on shore, fixing plant and processing rock, where they were as autonomous as their land-mining counterparts. After transporting the rock from lighters to a drying shed adjacent to an elevated railway and wharf, laborers then dumped the rock into a drying bin containing brick flues and perforated pipes. Finally, they loaded the dried rock onto oceangoing vessels. During a busy season, the company employed between five hundred and six hundred men.[46]

River mining paid well. Official statistics for the 1870s do not exist, but later records indicate that river miners could earn more than land miners, and both groups' earnings outpaced those of agricultural and textile laborers. Coosaw's skilled mechanics made $3.50 per day in 1880, compared to $2.50 at CMMC, and unskilled laborers at both companies earned a dollar, about twice as much as farm and mill workers. During 1890–92, engineers, foremen, and a tug captain for the river companies earned an average of $3.71 each day, but one garnered $6.33. By comparison, land foremen and engineers received $2.41.[47]

Management's perspectives of river miners were predictable. Coosaw Chief Manager Adger claimed that his company offered employment to previously "idle labor"—his label for black men who had left the plantations. Without Coosaw, Adger argued, "the negroes on those islands would either have starved or been thrown on the County for support, or lived by pillaging and robbery." He concluded that it was "gratifying to see the improvement in the healthful appearance, moral character and stability." They appeared "happy and contented," and never acted disorderly or insubordinate. Working more closely with the miners, David Lopez described them as "a class of laborers equal to any in the world."[48] Of course, context is important. Adger and Lopez were promoting Coosaw's claims for exclusive rights to state officials, and they knew that firm paternalism and industrious black labor scored points in Columbia.

Dangerous yet profitable for its workers, the river-mining industry made a strong start in its first five years, with production increasing each year, but it was not an immediately remunerative business. Marine mined twenty thousand tons during its first two years, grossing about half of what it spent on plant and equipment; after royalty and labor costs, the company was in a deep financial hole. Similarly, Coosaw spent vast amounts during this "experimental stage" and paid no dividends. Coosaw officials admitted that all their competitors (aside from Oak Point) were "financially weak," and their position was not much better. Despite paying no dividends, Marine's stock remained "in demand," because investors undoubtedly sensed the industry's enormous potential. A major source of financial uncertainty for all companies was the denial by Lloyd's of London to insure phosphate ships. Poor "stowage" techniques led to catastrophic cargo shifts clearing St. Helena Sound and other dangerous stretches; a third of the ships leaving Beaufort during 1870–73 were lost or damaged.[49]

South Carolina's Republican legislators crafted the early river-mining laws without the benefit of precedents and in a sea of partisan distrust and corruption. Predictably, the 1870 law formed the basis of enduring political conflicts and court cases within the industry. Accusations and justifications of monopoly would linger, as would arguments for large companies and concerns for common miners. Some merely sought wealth, but most black Republicans, carpetbaggers, Democratic businessmen, and other participants utilized ideological rhetoric to create and regulate an industry that they hoped would fix what ailed the state. Various entrepreneurs—domestic and foreign, southern and northern, and black and white—built the industry's foundations between 1869 and 1874. Overcoming chronic capital shortages endemic to the postwar South, owners, managers, and miners at Marine, Oak Point, and Coosaw assembled large, well-equipped companies that dominated the industry. Less-powerful businessmen fought for democratic access to the phosphate windfall by subcontracting hand miners. Antebellum ties to European and northeastern cotton markets facilitated South Carolina's entrance into world fertilizer markets.[50] Finally, black Sea Islanders used the river-mining industry to fashion a seasonal life around hunting, fishing, mining, and agricultural work. Land-based employees and independent rock gatherers enjoyed more autonomy than did boat crews. As integral to river mining as politicians and businessmen, the miners helped create the second part of an interrelated economic explosion in the lowcountry stemming from the 1867 "discovery" of phosphate rock. The burgeoning fertilizer-manufacturing industry was the third and most lasting piece of that explosion.

5

Convergence and the Fertilizer Industry, 1868–1884

Surveying South Carolina's fertilizer industry in 1882, Edward Willis wrote, "As the cotton mills are annually nearing the cotton fields, so the fertilizers based on crude phosphate rock must come to the fountainhead of the industry."[1] The convergence of the land-mining, river-mining, and fertilizer-manufacturing industries had a profound impact on Charleston, South Carolina, southern agriculture, and America's fertilizer industry. Beginning in Charleston and Beaufort, the lowcountry's fertilizer industry included local, regional, northern, and European entrepreneurs who sought to use the abundant nearby rock deposits to increase production, lower prices, and overcome southern farmers' skepticism of commercial fertilizers. Instead of shipping rock north and commercial fertilizer south, the new firms sought to mine and manufacture locally and use the savings to compete with northern producers.

South Carolina's fertilizer industry was born during a transitional period for regulators, farmers, and businessmen in the South. In contrast to the North, southern governments did not regulate the industry until the early 1870s. South Carolina's first fertilizer inspector, Charles U. Shepard Jr., merely listed local brands produced during 1870. A more stringent 1872 state law required detailed labeling on fertilizers. Another law empowered phosphate inspector Otto A. Moses to evaluate fertilizers' worth based on ingredients. Other states strengthened fertilizer laws and published chemical analyses, gradually but effectively limiting fraud in the South.[2] Increasing regulation in South Carolina also led to the growing importance of lawyers, and lawsuits over river mining and environmental pollution had significant consequences for fertilizer companies.

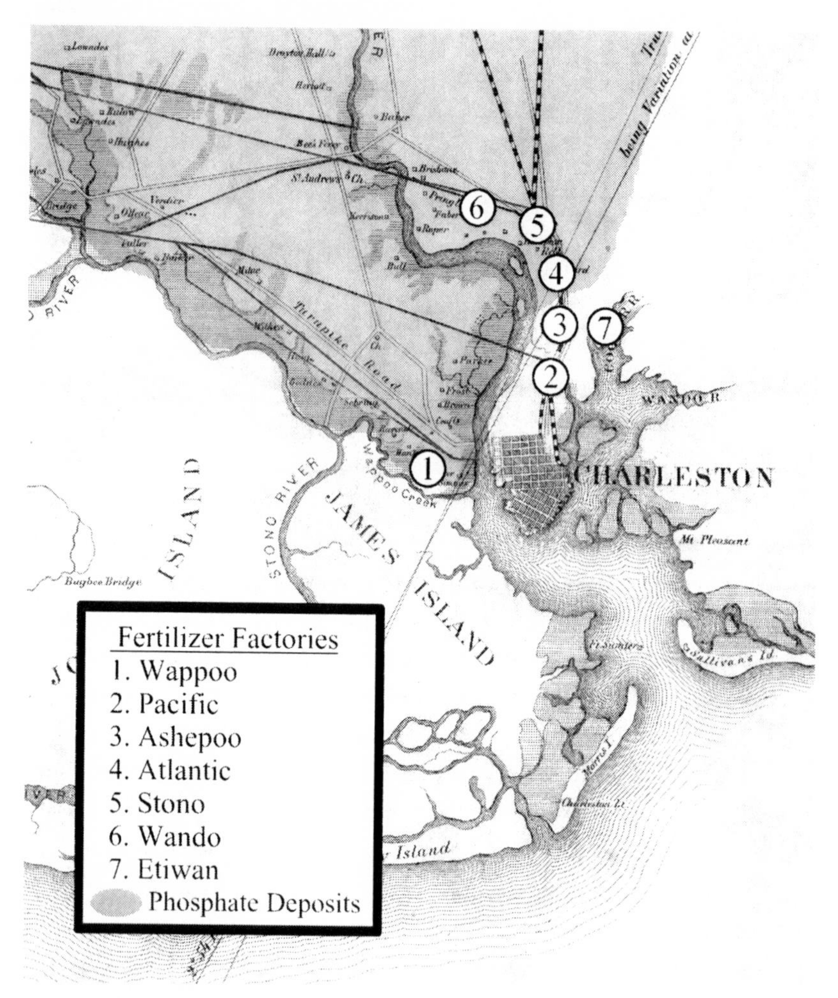

Map 4. Charleston-area fertilizer factories. Map created by author.

Entrepreneurs formed at least sixteen fertilizer companies near Charleston between 1867 and 1884. Some companies maintained small factories in the city, but the most successful ones established sprawling works on the Charleston Neck, a sparsely settled area north of the city.[3] Fertilizer factory mania soon spread to Beaufort, the center of river mining. Sea Island Chemical (Beaufort), Walton, Whann & Company (Wilmington, Delaware), Baldwin & Company (Savannah), and Hume Brothers & Company (Edinburgh, Scotland) established factories in Beaufort and used locally mined rock. Beaufort-area firms exported 75 percent of their product (mainly to England and Germany) in 1884, whereas Charleston firms exported only 5–8 percent. By the mid-1880s, manufacturing began in upstate towns, including Columbia, Chester, Anderson, and Greenville, but Charleston remained the state's center of production.[4]

Eight Charleston firms—Wando Mining and Manufacturing, Sulphuric Acid and Super-Phosphate (SASP), Carolina Fertilizer, Stono Phosphate, Atlantic Phosphate, Pacific Guano, and John B. Sardy's Wappoo and Ashepoo ventures—provide salient examples with which to study the companies' origins, plant, marketing, workers, and legal problems.[5] Such analysis yields an understanding of the strategies pursued and problems encountered by those who worked for and owned the companies. It also denotes an impressive degree of industrialization in the Charleston area and a greater variety of fertilizers available to southern consumers.

Origins

South Carolina's first fertilizer companies included entrepreneurs, capital, machines, and methods from the North and the South. Officers, board members, and stockholders were predominantly southerners, but much of the capital came from the North. Following the trend in river and land mining, fertilizer company offices were often mere nameplates, and their small bureaucracies—officers, a board of directors, a few managers, and a chemist—met in their agent's or another board member's office. Mirroring other Charleston businesses, many fertilizer firms were extensions of family commercial dynasties. For example, James S. Gibbes and Lewis R. Gibbes initially headed Stono Phosphate, but members of the Ravenel family, including President William Ravenel and Chemist St. Julien Ravenel, soon took control.[6] Occasionally, charges of nepotism caused disputes within these family-oriented firms. In 1878, Treasurer Arthur M. Huger complained to Stono's shareholders that Ravenel-family domination was behind a disastrous change in fertilizer formula involving Chemist Ravenel's marl works. President Ravenel quelled the mutiny, but the fray led to alterations in Stono's by-laws, giving the president and board "paramount" authority.[7]

Local factors, merchants, and lawyers dominated leadership positions in Charleston's fertilizer companies, and many of the city's larger businesses shared directors. In 1869, Wando's president and general agent and several directors headed commission-merchant and cotton-factoring firms, and three of them sat on boards of prominent local companies.[8] In the 1870s, Wando stockholders included prominent Charleston lawyers, several of whom were active in the Democratic Party. Forming a postwar "strategic elite," Broad Street lawyers replaced planters as the city's powerbrokers, especially in the three phosphate-related industries.[9]

The leaders' social status and business acumen were crucial for the success of early fertilizer companies. Prominent former Confederates with elite family names organized Atlantic Phosphate in May 1870. St. Julien Ravenel, collaborator on the CSS *David,* was the company's chemist, and Francis J. Porcher, a signer of the Ordinance of Secession, served as president. Treasurer Francis J. Pelzer, however, was the power behind the throne, and his factorage firm, Pelzer, Rodgers & Company, was the company's agent. Building what would become an upcountry textile empire, Pelzer formally took charge of Atlantic in 1872 and installed his partner as treasurer. Not surprisingly, Atlantic prospered, reportedly selling more fertilizer from 1872 to 1875 than any other local manufacturer.[10]

Officers and board members usually enjoyed the confidence of the stockholders, who typically gave them free reign in managing company policy. Exceptions did occur, however, especially when stockholders perceived that officers made bad decisions. Beginning in 1871, Wando's board borrowed "large sums" as the company expanded, and creditors accused the company of "watering the stock" during the 1870s. At the same time, President John R. Dukes, likely part of general agent William C. Dukes & Co., went bankrupt, a potentially crippling blow to the company's finances.[11] Wando survived the crisis, but it would not be the last fertilizer company to have its future clouded by its general agent's failures.

It was not long after the phosphate rock "discovery" that northern fertilizer firms, in search of raw materials and new consumers, joined the local companies in manufacturing phosphate-based fertilizer. The northerners were aware of their status as outsiders and used various methods to defuse southern hostility. With its guano supply dwindling, Boston's Pacific Guano leased part of Chisolm's Island in 1869 in order, initially, to supply its Woods Hole, Massachusetts, plant. Sharing the island with the Coosaw

Mining Company, Pacific Guano's deposit was, according to Harvard geologist Nathaniel S. Shaler, "the greatest development yet discovered." The yield soon convinced the company's leaders to manufacture fertilizer in South Carolina. Lacking Wando's local clout, Pacific Guano hired St. Julien Ravenel to be its chemist and planned a new factory in Charleston.[12]

New York guano merchant and former Pacific Guano agent John B. Sardy sensed the same opportunities in the lowcountry. During 1868–69, he purchased the Wappoo Saw Mill, converted it into a fertilizer factory, and stocked it with phosphate from his 1,600-acre Ashepoo Mines. The mill property included fifteen to twenty acres, was situated across the Ashley River from Charleston, and was accessible to land, rail, and sea transportation.[13] Like Pacific Guano, Sardy was well positioned to capitalize on the convergence of mining and manufacturing. In local advertisements, the New Yorker emphasized the southern origins of the fertilizer's raw material. And, just as Pacific Guano hired Ravenel, Sardy employed G. A. Trenholm & Son as general agents, hoping that the Trenholm name neutralized, in southern consumers' minds, his company's Yankee origins.[14]

Other northerners sought to grab southern market share by blurring the lines between manufacturer, brand name, and dealer. William L. Bradley (Bradley Fertilizer Company) manufactured fertilizer in Charleston but sold it through George W. Williams & Company under the Carolina Fertilizer moniker. Thus, in every ad, farmers were led to believe that Carolina was the locally made product of the leading Charleston businessman George W. Williams. The arrangement began in February 1869, but unlike other components of Williams' empire, the product was a small part of the Charleston market; all but one company outproduced Carolina during 1869–72. Such brand confusion was possible because press reports and industry literature often failed to distinguish among fertilizer dealers, manufacturers, and rock miners.[15]

Southern entrepreneurs responded to the northern invasion by exploiting the North-South cultural divide and cultivating newspaper reporters. Local companies advertised the "southern" qualities of their management and fertilizer brands to garner regional loyalty. Charleston's press was only too eager to assist. Touting the new ventures in terms of southern patriotism and local boosterism, *Courier* writers held up southern industrialization

The "CAROLINA FERTILIZER" is made from the Phosphates of South Carolina, and is pronounced by various Chemists one of the best Manures known, only inferior to Peruvian Guano in its Fertilizing Properties. These Phosphates are fully described in this volume, and possess qualities of the greatest value to the agriculturist.

We annex the analysis of Professor Shepard:

LABORATORY OF THE MEDICAL COLLEGE OF SOUTH CAROLINA.

Analysis of a sample of CAROLINA FERTILIZER, personally inspected.

Moisture expelled at 212° F		16.70
Organic Matter, with some water of combination expelled at low red heat		16.50
Fixed Ingredients		66.80
Ammonia		2.60
Phosphoric Acid—Soluble	6.96	Equivalent to 11.27 Soluble Phosphate of Lime
Insoluble	6.17	Equivalent to 13.48 Insoluble (bone).
	13.13	24.75 Phosphate of Lime.
Sulphuric Acid	11.01	Equivalent to 23.65 Sulphate of Lime.
Sulphate of Potash		80
Sulphate of Soda		3.50
Sand		11.06

On the strength of these results I am glad to certify to the superiority of the CAROLINA FERTILIZER examined.

C. U. SHEPARD, Jr.

GEO. W. WILLIAMS & CO., Factors, Charleston, S. C.

Figure 10. Innovative advertising, Carolina Fertilizer, 1870. By permission of South Caroliniana Library, University of South Carolina, Columbia.

as a solution to what planter elites labeled the "southern labor problem"; modern agriculture and factories might resurrect the southern economy and end elites' dependence on freedpeople.

SASP was one of several companies the local press doted upon with bursts of southern devotion and lowcountry promotion. Not satisfied for his city and region to merely supply raw materials and to "yield to the Northern manufacturers the lion's share" of fertilizer sales, an editor declared that the success of local companies such as SASP and Wando was an issue in which "our whole community is interested." Skilled white "mechanics and tradesmen" found jobs, and farmers were spared long-distance freight expenses. Especially noteworthy to the *Courier*'s conservative readership, cheaper fertilizer allowed farmers to increase production while employing fewer allegedly troublesome and expensive black laborers. Another writer argued that SASP was proof that "our people are becoming awake" to new opportunities after the war and that prosperity was "steadily growing" in South Carolina. "Let our people leave politics to politicians and demagogues," he declared, "and enter upon the work of real, substantial reconstruction of our fortunes."[16] These words reflected the bitter resignation of many southern whites during Reconstruction, who felt that economic gain was their only remaining forum for independence.

That SASP was likely linked to the Philadelphia-dominated mining behemoth Charleston Mining and Manufacturing Company (CMMC) was not important to local boosters. The *Courier*'s editorial staff approved of SASP's founders, including C. G. Memminger, Robert Adger, and Nathaniel A. Pratt. Memminger and Pratt's Confederate service gave them the ultimate southern credibility, and Adger's family business was well respected.[17] Blessed by the *Courier*, SASP's leadership represented some of antebellum Charleston's brightest commercial leaders and conservatives, who could be trusted, according to the *Courier*, to lead the city toward what elites considered a sensible New South.

SASP began advertising in late December 1868 and immediately sought to capitalize on its perceived strengths. "Under the direction entirely of Southern men of high character," the new company boasted about its fertilizer made at "home" and supplied by "near by" phosphate mines. Company advertisements introduced fertilizers with the local "Etiwan" brand and prominently displayed Memminger and Pratt as leaders.[18] Ignoring

Wando's fertilizers, SASP tried to position itself as the only patriotic, high-quality choice for the region's farmers.

Industry leaders led their companies to profitability during the first few decades in business. Wando rapidly expanded its annual fertilizer production from three tons in 1867 to 8,400 in 1870. Despite the Dukes crisis, Wando continued as a stable and profitable company, earning as much as 30 percent profits during its first decade.[19] One of the major reasons that Charleston-area firms prospered was that most realized that fertilizer manufacture should not be a small-scale, urban-centered operation. New South industrialization called for dramatic changes in scale.

Plant

Eight of Charleston's leading fertilizer manufacturers began production between 1867 and 1871 in or near Charleston. All purchased local phosphate rock for their manufacturing, although some also mined for a few years. Merging traditional milling techniques with increasingly sophisticated mechanical technologies, superintendents and managers sought to improve their manufacturing methods. Many of the improvements included labor-saving machinery and steam engines. Managers purchased existing factories, constructed new factories at wharves in the city limits, or started from scratch in green fields on the Neck. The most successful companies recognized that creating manufacturing complexes north of the city was the way of the future. The Neck offered ample space for plant expansion and opportunities to fully utilize rail and river transportation. The Neck also offered managers a place to manufacture using dangerous and foul-smelling chemicals without fielding complaints from city residents. One such chemical was sulfuric acid, necessary to dissolve phosphate rock for fertilizer and another important milestone for lowcountry companies as they evolved from local shops to national producers.[20]

As the first postwar fertilizer manufacturer, Wando built in late 1868 its second city factory (a four-story, barnlike structure), a warehouse, and a drying kiln next to the Cooper River. With political and economic conditions unsettled and the fertilizer business unproven in the South, Wando's choice of location reflected a safe if not forward-thinking choice by its founders and investors. But manufacturing in the city had several drawbacks. Plant expansion was difficult and expensive in the crowded city, where land and buildings were at a premium. The Cooper side of the peninsula featured good docks, but most local rock came from the Ashley River, and most companies shipped their fertilizer by rail. The city was poorly served by rail. Antebellum Charleston's leaders stopped the rails approximately eighteen blocks north of the commercial district's center for fear of noise and fires. Until the 1880s, all freight had to be transported by wagons to the docks, adding to costs.[21]

More drawbacks included residential complaints and lawsuits about the smells and noxious fumes associated with fertilizer manufacturing. Most of the ingredients in fertilizers smelled bad and some, especially sulfuric acid, were dangerous. In 1869, a Charleston resident complained to the Mayor's Court that "guano" (fertilizer) stored near the South Carolina Railroad (SCRR) freight depot was "offensive to the neighborhood." Mayor Gilbert Pillsbury dismissed the case, but the resident planned to pursue the case in the county court system. Hoping to forestall further legal action, an officer of the unnamed company declared that the firm planned to build a storage shed in a "less populous portion of the city," likely on the Neck.[22]

Fertilizer manufacturers soon realized that the Neck's green fields offered abundant and inexpensive real estate. Companies could build large industrial enterprises, including multiple docks and rail spurs, confident that local and county officials would ignore complaints by working-class residents. The narrowest part of the peninsula that formed the city, the Neck consisted of farms, marshy areas, and cheap housing. During the war, poor residents continued to move north into the Neck, but contemporary maps from 1863 and 1872 showed few roads or buildings. The Neck's geography offered advantages for companies seeking to integrate several large buildings with railroad and shipping. The western side offered large open spaces adjacent to the Ashley, with few areas impeded by marsh; the eastern side featured similar access to land, rail, and river (the Cooper and its tributaries). The narrow land also funneled the SCRR and Northeastern Railroad close to the rivers, decreasing the distance of each fertilizer company's railroad spurs.[23] Factories in need of supplies still benefited from their proximity to Charleston. In short, the Neck offered manufacturers advantages that the city could not match.

Although most fertilizer companies established their new facilities on

Figure 11. Etiwan fertilizer factory, Charleston, 1870. By permission of South Caroliniana Library, University of South Carolina, Columbia.

the western side, the first corporate settler on the Neck chose the eastern side. SASP's board purchased land near the old "Navy Yard" on Shipyard Creek and planned a test farm and a large manufacturing complex. Pratt used the farm—an example of SASP's forward-looking techniques in research, development, and marketing—to test new formulas and woo discerning farmers. Construction began in September 1868 on the "Etiwan Works"—the main building of which was over 50 percent larger than Wando's city factory—at a location that enjoyed deepwater access to the Cooper, Charleston Harbor, and the Atlantic. Their choice of location likely reflected the northern orientation of its investors but did not create a disadvantage in accessing local phosphate rock (likely from CMMC); the well-financed company merely moved it from the Ashley by rail and creek.[24] By the early 1880s, the Etiwan Works' buildings covered four and a half acres and included a factory and at least seven smaller buildings, sheds, and offices. It employed between sixty and one hundred hands and produced twenty thousand tons of fertilizer per year.[25]

The Wando, Pacific Guano, Atlantic, Stono, and Ashepoo companies followed SASP to the Neck during the next decade, and the team of St. Julien Ravenel and David C. Ebaugh designed and built many of the plants. Southernmost on the western Neck, Pacific Guano's Rikersville complex emerged in 1869 as a worthy competitor to SASP's. Chemist Ravenel and

Superintendent Ebaugh created and managed a complex accessible to the Ashley and railroads; it included a bone mill, fertilizer factory, storehouse, and acid chambers. By 1875, its Woods Hole and Rikersville factories employed between four hundred and five hundred men and had fertilizer sales of over thirty-five thousand tons.[26]

Unlike SASP and Pacific Guano, Wando migrated to the Neck in stages. When the Ebaugh-built city factory and warehouse proved insufficient, the company built adjacent to its Bee's Ferry mines in 1870 a processing plant that included a rail spur from the mines, rock washer, sulfuric-acid chamber, and wharf. Finally, around 1880, Wando erected a large complex on the northwestern edge of the Neck that replaced the city and mine plants. Employing one hundred hands, Wando's "new works" was an industrial complex spread over many acres, incorporating fifteen buildings and transportation facilities, and still vertically integrated with their nearby mines. Easily dwarfing the old city factory (30 × 300 feet), the new complex included a factory (80 × 50), storehouse (50 × 300), warehouse (40 × 150), acid chamber (70 × 226), and shipping shed (28 × 335).[27]

Following the herd, Atlantic and Stono hired Ravenel and Ebaugh to build their similar plants on the Neck in 1870–71. Atlantic's main factory building was smaller than SASP's, but it expanded its plant throughout the next decade. By the mid-1880s, Superintendent A. M. Rhett's one hundred Atlantic employees worked in seventeen buildings on a ninety-acre industrial manufacturing complex, producing up to thirty thousand tons of fertilizer annually. Stono's board purchased a 180-acre farm just north of Atlantic's property that featured ample Ashley frontage.[28] Roughly equal in size to Atlantic in terms of capital invested and hands employed, Stono's complex in 1880 produced more fertilizer than any local competitor and kept pace with Etiwan in new technology. Like Wando, Stono was vertically integrated, mining the rivers but also buying local rock for its plant.[29]

Another fertilizer manufacturer who mined rock, John B. Sardy delayed moving to the Neck because his Wappoo factory and Ashepoo Mines functioned well together. Without an acid chamber, however, he was forced to import expensive sulfuric acid from the North. Seeing the larger firms prosper on the Neck convinced Sardy to form the Ashepoo Mining and Manufacturing Guano Company in 1874 and build a new complex on forty acres between Pacific Guano and Atlantic. The new main building was

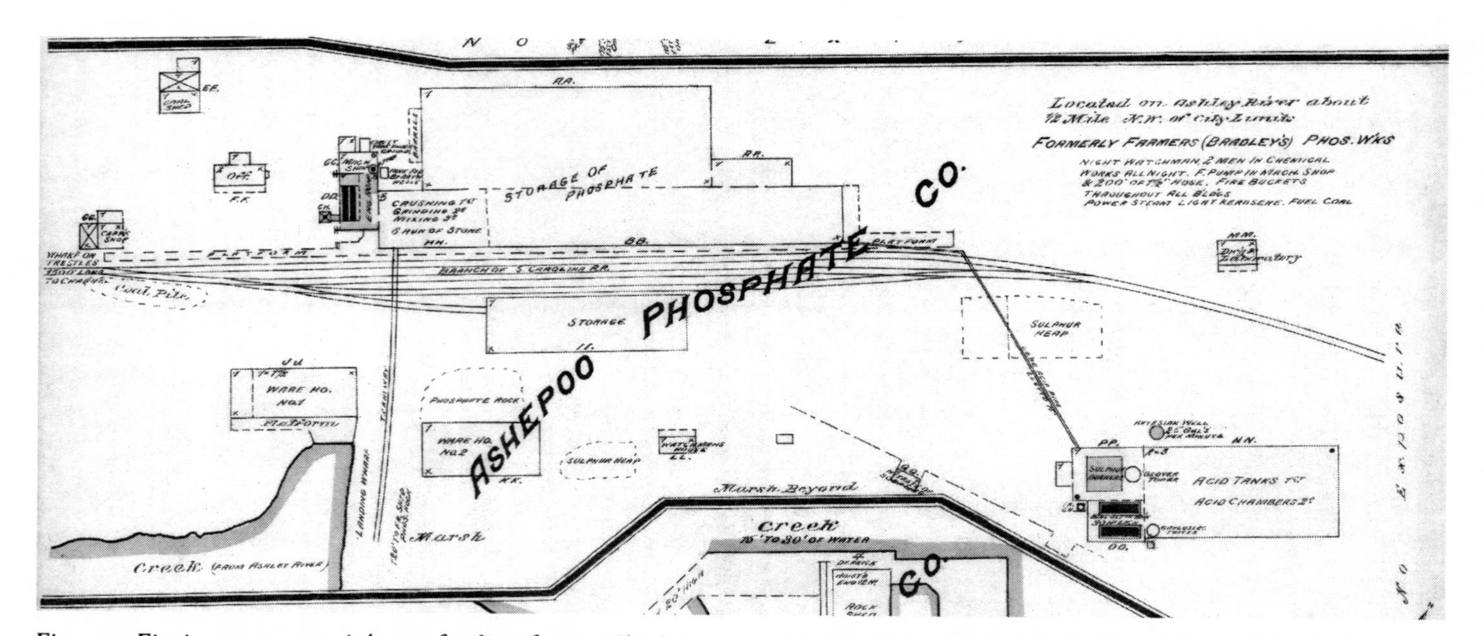

Figure 12. Fire insurance map, Ashepoo fertilizer factory, Charleston, 1884. By permission of South Caroliniana Library, University of South Carolina, Columbia.

similar in size to Atlantic's, but Sardy kept Wappoo operating as well. Both factories continued producing fertilizers for many years, but they fulfilled vastly different roles. An old factory, Wappoo remained a relatively small processing plant for grinding phosphate rock. In contrast, Ashepoo was a modern manufacturing complex that continued to expand.[30] The factories represented the two roads faced by the leaders of Charleston's fertilizer companies in the 1870s, that of the past and the future. By 1883, the Neck was a busy industrial district, and the plants of forward-looking companies dominated the landscape.

Carolina Fertilizer never established a manufacturing complex on the Neck. Growing competition from Etiwan and the other large companies may have forced Bradley and Williams to choose between moving to the Neck or stop producing altogether. The brand disappeared around 1880. A small local manufacturer throughout the 1870s, Bradley chose to focus on mining and his other fertilizer ventures. Primarily a banker, Williams abandoned the diversions of river mining and the fertilizer trade.[31] Carolina's exit signaled the end of the small producer in the lowcountry.

The stampede to the Neck elicited more favorable press coverage in Charleston. While most residents welcomed the exit of the dangerous and smelly industry, the conservative *Courier* celebrated the lowcountry's expanding industrialization. The factories were now bigger, but the themes remained the same. Machines held the promise of solving the alleged labor problem. New southern investments in industry meant that the South, South Carolina, and Charleston could economically compete with, and perhaps whip, the North. Touring SASP's facilities, a *Courier* reporter was astonished that moving rock from dryer to crusher was the "only" manual operation in the "manipulation" process. "So complete and perfect is the machinery used," he continued, "that but ten hands are employed about the manufactory." Another reporter at Pacific Guano's factory marveled that "the work is done principally by machinery, and there is therefore little use for workmen." A third article in the *Courier* reflected a sense of defensive, almost bitter southern patriotism: Atlantic's new manufacturing facility was proof that the state had "more life than what carpers and jealous rivals suppose and admit."[32]

The reporters, however, were prone to exaggeration. The production process was not as automated as the *Courier* asserted, and there was plenty

Figure 13. Four scenes from a fertilizer factory, Charleston, 1880. By permission of South Caroliniana Library, University of South Carolina, Columbia.

of manual work for factory workers. A description of Wando's city factory illuminates the typical manufacturing process in 1869. Under the direction of Superintendent Thomas D. Dotterer, Wando's four-story city factory contained "all the modern appliances" for grinding, mixing, and packaging fertilizer. Thirty workers moved rock from the river wharf via handcars to "two large furnaces and ovens" that removed moisture from the porous rocks. The next steps involved grinding the rocks into powder. Factory hands pushed the handcars up inclines to the second-floor hopper and "cracking mills" and hoisted the smaller rocks to a third-floor machine that grinded them into powder. The final steps included drying, mixing, and bagging procedures. Laborers poured the powder into a fourth-floor drying hopper and then into a large (forty-ton daily capacity) "bowl" for additive mixing. Back on the third floor, laborers broke up the cooled hard mass with picks, passed the fertilizer through rollers to restore its powder form, and dried it again. Back on the first floor, workers filled and weighed 167-pound bags, and "a number of boys" sewed the bags shut. A "fine suction blower" allegedly removed fumes from the factory's floors, adding "much to the comfort of the operatives." Stacked in the "shipping department," bags of Wando Fertilizer were ready for sale. The process remained essentially the same in the 1870s and 1880s.[33]

Although SASP's shipping and manufacturing operations were nearly identical to Wando's, the Etiwan Works had bigger and better equipment, including a pair of steam engines to power machinery and elevators and a "disintegrator" machine to refine the newly mixed fertilizer from paste to powder. But both companies manufactured in a batch process, despite some resemblance to continuous flour production. Indeed, the new fertilizer factories purchased machines from the same companies that outfitted flour mills. For example, Pacific Guano bought its disintegrator from Holmes and Blachard of Boston, a flour-equipment supplier. Unable to make the process continuous, fertilizer companies raised production by improving transportation and employing larger mixers and curing dens to increase batch size.[34]

More significant than larger batches, the successful launching of SASP's sulfuric-acid chambers, Pratt later claimed, "revolutionized the fertilizer trade." Overcoming the skepticism of chemists and SASP board members who doubted that sulfuric acid could be manufactured in the Deep South, Pratt and his Etiwan Works became the first operation south of Baltimore to produce it on a commercial scale. Soon after, he expanded the chambers to create the nation's largest production facility, thus reducing supply costs and interruptions and taking full advantage of the nearby phosphate rock. Pratt proved that large-scale fertilizer production demanded fully integrated sulfuric-acid chambers. Not surprisingly, most of the other Charleston fertilizer manufacturers built acid chambers between 1870 and 1872.[35]

Marketing to Modernity

Besides creating cutting-edge industrial sites on the Neck, lowcountry fertilizer firms actively promoted their brand and products in increasingly modern ways. The new entrepreneurs noticed that the postwar business climate was moving away from intimate dealings among gentlemen and toward mass production and mass marketing across the South. Converting skeptical southern farmers into consumers by establishing brand names and courting favorable press coverage was new territory for Charleston's gentlemen-businessmen. Focused on product quality and pricing, farmers seemed to respond to flexible credit policies and positive crop results as much as to instructional advertising, leaders' reputations, quality assurances, scientific experts, and farmer testimonials. The early companies often relied on "agents" to market their products and interact with consumers. Agents were usually commission merchant firms whose principal often held the treasurer position in the fertilizer company. Agent weaknesses sometimes spread to the manufacturer, leading to changes in the marketing system. Combining the expansion of plant with effective marketing, boards led their fertilizer-manufacturing companies to modernity.

Agents were the initial marketers in the fertilizer industry. Although agents often held dominant positions in fertilizer companies' hierarchies during the first decade, the practice evolved as markets expanded during the 1870s and 1880s toward a more decentralized system of traveling agents or multiple agents in different towns. Large regional firms like SASP employed general agents who advised subagents. SASP's list of seventeen agents in towns across the Southeast, under the direction of general agent William C. Bee of Charleston, displayed the company's growing network in 1870. By 1871–72, the company had thirty subagents in South Carolina alone. Farmers' questions came to Bee, who then forwarded them to SASP's assistant chemist, W. W. Memminger.[36] National firms, such as Pacific Guano, employed a similar marketing structure. General agent John S. Reese & Company of Baltimore, and local agents (such as John N. Robson & Sons in South Carolina) under Reese's direction, sold Pacific Guano's products. When Reese replied to questions about the firm and its fertilizer, he emphasized cost, quality, and character. Reese praised Pacific Guano's leaders in terms of "respectability," "character," "reliability," and "best scientific ability," and repeated the phrase "public confidence" five times. Selling to farmers from New Jersey to Alabama, Pacific Guano manufactured, according to Reese, at the "lowest possible cost," a product of the "highest real excellence," in order to reap "small profits, large sales, and a permanence of trade."[37] Reese was the company's bridge from the era of personal transactions to a new era of mass marketing.

While Bee and Reese appear to have been reliable and efficient marketers for their brands, other companies became too dependent on general agents. Stono's reliance on the agent-treasurer system led to a "Stono

Rebellion" of sorts among stockholders before the 1878 annual meeting, highlighting the firm's flawed management structure and business culture. Paul C. Trenholm and other stockholders charged President William Ravenel's administration with "a want of energy and efficiency" in losing almost $135,000 from 1872 to 1878. Raising questions about oversight and chain of command, stockholders blamed Ravenel for not changing the sales system that was "so ruinous" to Stono's finances. Their accusations were on target. Earning commissions on sales, whether cash or credit, but not responsible for collection, Stono treasurer Joseph D. Aiken's Aiken & Company had the impetus to make as many sales as possible, even in risky credit situations.[38]

President of a mature but not profitable company, Ravenel pleaded impotency and patience. He claimed that Stono's earlier by-laws had allowed the treasurer to dominate the system and that since the board ran the company "without capital," it had to pay Aiken a "high" sales commission of 5 percent. The board replaced Aiken with W. B. Williams of Williams, Black, and Williams, but the company's finances continued to worsen. Ravenel and the board had begun a shakeup even before Trenholm's rebellion. Although the agent-treasurer system had been, according to Ravenel, regarded as "a *sine qua non* with Phosphate Companies," Stono's board abandoned the old system and transitioned to traveling agents, "a resort entirely new in the trade." With better regional coverage and salesmen responsible for completed credit sales, the new system promised better results. President Ravenel absorbed the title of treasurer and predicted renewed profitability.[39]

Possibly Charleston's first fertilizer company to abandon the old system, Stono was able to change because, after dropping Aiken, it lacked a strong tie to one commercial firm. While Pelzer, Rodgers & Company controlled Atlantic, none of the many factoring companies represented on Stono's board held a dominant position. More significantly, the shift to traveling salesmen was representative of bigger changes in Charleston and southern business.[40] Factoring, commission merchants, and importers served the Old South well but were ill suited to the advent of industrialization, mass marketing, and modern corporations. Charleston's fertilizer companies were entering an era in which advertising, brand name, and mass production and consumption were more significant than personal relationships.

Agents and boards of directors were innovative and evolving in targeting fertilizer consumers. An important lesson, especially as the industry transitioned from personal relationships to mass marketing, was that brands were important. For example, "Sulphuric Acid and Super-Phosphate" sounded impressively scientific and initially served the company well in highlighting Pratt's sulfuric-acid chambers. But the convoluted name ceased to be relevant by the mid-1870s, when most competitors had acid chambers. Farmers purchasing SASP's fertilizer were more likely to remember "Etiwan." A Native American name for the Cooper River, "Etiwan" reminded farmers that it was a southern company manufacturing with lowcountry phosphates. SASP's board changed the company name to the Etiwan Phosphate Company in January 1878.[41]

As cotton prices fell and commercial fertilizer consumption rose during the 1870s, boards and agents responded to farmer complaints about high fertilizer prices by adding more product and price choices. The Rikersville factory manufactured "Soluble Pacific Guano" as well as "Compound Acid Phosphate of Lime," a cheaper alternative to, but supposedly "as valuable a fertilizer as," the more expensive flagship brand. Etiwan sold three types of fertilizer—Etiwan Guano, Etiwan Crop Food, and Etiwan Dissolved Bone—and pulverized phosphate rock, Etiwan Ground Bone. Aside from the fact that Etiwan Guano contained no guano and Dissolved Bone and Ground Bone had no bone, the company seemed to offer a full menu of fertilizers for discerning farmers. Not to be outdone, Sardy produced five different types and sold other brands as well. Prices and recommended dosages varied. Sardy sold Ammoniated Soluble Pacific for forty-five dollars per ton and recommended 250 to 350 pounds per acre. Acid Bone Phosphate cost twenty dollars less but required three hundred to four hundred pounds. Ground Carolina Phosphate (lacking additives) was Sardy's least expensive product and sold for fifteen dollars.[42]

Fertilizer firms and agents offered flexible payment options. Sardy accepted cash or added five dollars to the price of Ammoniated Soluble Pacific for credit sales. Williams & Company sold fertilizers, including the Carolina brand, "on time," adding 7 to 10 percent interest. The company also accepted the "cotton option" as payment from cash-poor farmers, but this became controversial. Citing falling cotton prices in 1879, Williams and other merchants had increased by 16 percent the amount of cotton needed to pay for each ton of fertilizer. Already squeezed by high prices, farmers accused Williams of slyly using the cotton option to increase fertilizer prices.[43]

Agents like Williams hoped that advertising would overcome farmer distrust of fertilizer pricing, value, and purity, but in the transition from personal to mass marketing, there were no easy answers. Agents tried a variety of advertising techniques, including reasoning, eye-catching logos, farmer testimonials, appeals to personal goodwill and patriotism, and scientific expertise. Stono stocked their almanacs with treatises on soil preparation and cost analyses of "manuring." Williams preferred reasoning with farmers, but he was not above using gimmicks to grab their attention. His ads carefully calculated the Carolina fertilizer used per acre, credit and cash prices, and estimated profits. Carolina ads also featured vivid images of dinosaurs decaying into fossils and the misleading caption "These phosphates are the remains of extinct land and sea animals." Williams learned that reason only goes so far in marketing, but his ads, like SASP's product names, also sought to reassure farmers of the organic nature of their products.[44]

Quality was also crucial. The skeptical farmer had to be convinced that the product contained the stated ingredients and would perform as advertised. During the transition to effective state and federal fertilizer monitoring and to mass marketing, farmers still preferred an element of personal trust in their transactions. Agents used testimonials to bridge the trust gap with the consumer. Pacific Guano brochures in the South Carolina market featured letters written by fellow southerners. Farmer George Jones of Fruit Hill wrote that after using Soluble Pacific Guano, "the prospect is fine for an elegant crop" and that compared to other fertilizers, the brand was "as good, if not superior." It seemed to "hold out and not exhaust."[45]

Another marketing tactic was to appeal to an older, more personal way of doing business, thus allowing the company to portray itself as an old friend from the country, not a new stranger from the city. Stono's 1880 Almanac included references to family, friends, children, church, honesty, thrift, and "personal independence" and also to "pleasant relations" and "mutual friendship and confidence" among the company, its agents, and farmers. "Money is a good thing no doubt," the Almanac declared, "but we think that warm sympathies and kindly affections have their value also."[46]

Especially during Reconstruction, effective marketing in the South included appeals to patriotism. Several firms sought to portray successful southern businesses as vehicles to economic, if not political, independence from the North and other "foreign" countries. Stono's Almanacs reminded southern planters that instead of paying Peru for overpriced guano, they could, with help from Stono, produce an equally effective product at home. Taking on the North as well, Stono demanded, "Why should Connecticut furnish the South with hay and onions, while any good Southern farmer could make a very respectable living" from growing the same crops in addition to cotton? Ample use of Stono's fertilizers made sense considering "the present condition of the labor" in the South.[47]

In the rural South, science was a double-edged sword for marketers; farmers often complained that "book farming" had no basis in reality, but they also, grudgingly, admitted that scientific research could improve crop yields. Fertilizer companies and their agents, therefore, had to balance claims of scientific expertise with evidence of practical experience. SASP's aggressive marketing efforts included promoting itself as the expert in matching fertilizer to southern farmers' needs. Its test farm not only drove research and development, but it also served to convince farmers that SASP perfected its products under real conditions. Aside from *Courier* advertisements, the company sent out "circulars" describing the science behind, and practical uses of, the company's different fertilizers. Sporting the slogan "Feed Your Land and It Will Feed You" on its cover, SASP's *Almanac 1872* featured a "Comparison of Advantages" between each grade of fertilizer, analysis of fertilizing profitability, and detailed "Directions" for fertilizing specific crops. Typical during the decade in which most southeastern farmers converted to the gospel of fertilizer, SASP's instructions formed an important bridge to the consumer.[48]

Each company featured "chemists" in ads to demonstrate their claims of scientific expertise and to market their fertilizers. St. Julien Ravenel was in high demand because of his scientific reputation, his Confederate service, and his well-known lowcountry planting family. Charles U. Shepard Jr.'s credentials, as well as his father's reputation, made his certification of various fertilizers equally valuable. The son of C. G. Memminger, SASP's W. W. Memminger raised the company's visibility within the state's agricultural journal, the *Rural Carolinian*. Memminger eagerly responded to "practical farmer" Edward E. Evans' letter praising SASP's "excellent" results in what amounted to an extended advertisement for SASP. The chemist's intelligent response also underscored the fact that Etiwan fertilizers came from a marriage of scientific authority and practical results. In another article, Memminger touted SASP ("among the foremost in the world"), its machines and sulfuric-acid facilities, and its laboratory ("among the most complete in the South").[49]

Workers

While agents and board members aggressively pursued the southern consumer, they and their managers struggled to secure a reliable workforce and reacted to factory hands' demands. Managers ignored workers in written documents, but newspaper coverage of an 1873 strike illuminates the concerns and standing of fertilizer laborers. In an instance of labor unity and a sign of industrial consciousness, black fertilizer workers joined others in demanding higher wages, but in their case, wages were directly related to their labor with chemicals. In addition, early fertilizer-mill employees, black and white, faced dangerous working conditions and challenges finding suitable lodging in the still-rural Charleston Neck, an area that experienced substantial changes between 1870 and 1880. Together, workers' demands and employers' responses offer a glimpse into the heart of the fertilizer-manufacturing industry.

Little information exists about early fertilizer workers, due to unclear terminology in the 1870 census and the paucity of company records, but as always, race mattered in Charleston. Following hiring practices at local mining firms, such as board member Robert Adger's Coosaw Mining Company, SASP managers likely hired a few skilled whites and numerous unskilled blacks. Despite working on the Neck, black fertilizer workers were part of Charleston's working classes. Following a largely successful strike by longshoremen in early September 1873, fifty of the city's black laborers declared a "general strike" to achieve a $2.50 minimum daily wage. The group then forced the closing of "most" of the city's factories and extended the strike to the Neck's fertilizer works. Sixty black hands from the Etiwan Works walked off the job on September 11, initially demanding the $2.50 minimum but later lowering that to $2.00. SASP Superintendent Caspar A. Chisolm refused to raise wages and closed the Etiwan Works, retaining only the white laborers at the sulfuric-acid chambers. The sixty

SASP strikers then attempted to gain access to Atlantic and halt production but were unsuccessful, presumably stopped by managers. Workers at neighboring Pacific Guano struck on the same day and, together with SASP laborers, subsequently forced workers at Atlantic and Stono to join the strike.[50]

The fertilizer strikers were not organized, unified, or well positioned. The *Courier* reported that "dissatisfaction" and "the wish to return to work" were widespread and that factory owners were confident of replacing strikers as long as "the violence of the mob" could be held in check. Although *Courier* coverage could be dismissed as conservative spin, economic realities and other evidence reinforce the newspaper's view. Ample unskilled labor was available from rural blacks migrating to the city. A published appeal on September 15 by "Many Laborers" was revealing: "At a large gathering of the laborers of the Phosphate Mills we expressed our willingness to inform the public as to the causes of our strike. We cannot meet our expenses with five dollars or six dollars a week. We have been striving to live by it for the past three or four years. The work requires more clothes and shoes than any other business in the city; and these are the causes and reasons of the strike." The letter provided evidence that their work exposed fertilizer workers to highly corrosive chemicals. Rather than striking for safer working conditions, they demanded higher wages to replace their tattered clothing. The appeal was their only sign of unity and the strike's last gasp. Workers began returning to the fertilizer works the following day.[51]

The *Courier* portrayed the strike as a week of violence toward white businessmen and black workers "directed by no organized society at all, but simply by about fifty negroes, who have armed themselves and forced the other laborers into a strike." The newspaper described the ringleaders as "twenty men and as many vagabond boys—that class of boys who have been at the bottom of every riot that has ever occurred in Charleston."[52] Violent bullies may have been catalysts, but the newspaper's coverage of fertilizer hands contradicted its theory on the strike's leadership. SASP workers, not vagabonds, led the strike on the Neck, and harsh working conditions at fertilizer works motivated them.

Success with strikes, whether in Charleston or other southern cities, depended on race, skills, numbers, and organization. Local strikes in 1869 by the Longshoremen's Protective Union Association and the Journeymen

Tailors' Union were successful because both unions were made up of skilled workers who dominated their respective industries. Other strikes in the South begun by skilled union workers, such as Mobile's dockworkers in 1867 and New Orleans' black longshoremen in 1872, tended to spread to unskilled workers, but race and skill were key factors when replacements threatened to take strikers' jobs. Charleston's black and white painters struck in 1869, but blacks found themselves deserted when whites received higher wages.[53] The earlier strikes should have served as cautionary tales for the unorganized, low-skill black fertilizer laborers working with whites. But their willingness to strike and demand economic justice indicates that they were not docile on the factory floor and that they were maturing as industrial workers.

Census, industry, and company records fail to reveal daily fertilizer wages in the early 1870s, and it is unclear if wages over the next decade increased due to the strike or decreased after 1873 as a result of the economic downturn. There is no evidence of another fertilizer strike during the decade. The better available data for 1880 shows that wages in the lowcountry's fertilizer industry were comparatively good. Census enumerators listed sixty to one hundred adult males working ten-hour days at the Etiwan Works, with skilled hands making $2–2.50 per day and the unskilled earning a dollar. By comparison, upstate textile workers averaged fifty cents in 1883, and skilled weavers earned ninety-eight cents.[54] Favorable and steady pay for partly indoor work at profitable companies during otherwise lean economic times likely explains the lack of strike activity after 1873.

Dangerous chemicals were not the only hazards fertilizer mill hands faced. Industrial work was inherently unsafe in many nineteenth-century factories, and workers often faced the choice of work or safety. When workplace injuries occurred, some workers sued their employers for relief, but given employers' superior legal resources and the pro-business legal atmosphere, it is not surprising that little evidence of such lawsuits exists. Robert B. S. Sanders sustained serious injuries from a fall while working as a carpenter at the Etiwan Works in 1881 and sued the company for ten thousand dollars. Etiwan's lawyer, Augustine T. Smythe, stated that such an accident was a "risk of which he took upon himself when entering into such employment." He also argued that Sanders' negligence caused the accident. Although several of Sanders' coworkers testified on his behalf, the judge agreed with Smythe and dismissed the case, citing the plaintiff's failure to prove the company's negligence.[55]

Another concern for early fertilizer workers was housing. In 1870, the Neck was a rural farming community, not ideal for housing an industrial workforce. Before the strike, company leaders planning factories realized the need to accommodate white workers on the Neck. A few houses would suffice; compared to the later textile industry, the fertilizer industry had fewer white workers, and therefore a mill village was not necessary. According to President James S. Gibbes, Stono supplied "a suitable superintendent's house, dwellings for the watchman, engineers' and laborers' houses, stables, &c." during the initial building phase, 1870–71. By ignoring race, Gibbes likely signaled that the houses were for whites. Using code words to similarly indicate a whites-only policy, W. W. Memminger explained in early 1873 that SASP provided housing "so that at all times the property of the Company is protected by the presence of a large number of intelligent and efficient men." The security force included seven white employees—the superintendent, engineer, and sulfur burners—and their families in four houses. Memminger implied that the company needed protection from the unintelligent and inefficient—white euphemisms for freedpeople. Months before the strike, he made no mention of housing black workers, the vast majority of his 120-man workforce. Interestingly, Memminger anticipated the strikers' concerns, assuring stockholders that the only dangerous material at the Etiwan Works was nitrate of soda, safely stored in a fireproof brick magazine.[56]

In the decade following the strike, Charleston's black and white fertilizer workers continued settling the Neck. Stono appears to have supplied "cottages" for some of its workers as late as 1875, but there was no evidence of group housing by 1880. Although terminology confusion and nearby mines render distinguishing factory workers from miners an inexact science, 1880 census population schedules reveal many black and white workers living near the factories. The portion of St. Phillip's Parish on the Neck included eighty-eight fertilizer workers, including seventy-one black men, living among the parish's 295 mostly single-family dwellings. Etiwan's "local superintendent," Scotland-born John Duncan (white), lived with his family and servants near the factory, and his house lay between four black families and three white families, each of whom included phosphate mill workers.

The parish's fertilizer workers were a stable workforce, family men with the same average age as their land-mining counterparts. Almost two-thirds had children, and three-quarters were married to wives "keeping house." The demographics reflect the increasingly mature and profitable status of the local firms.[57]

Legal Issues

While successfully navigating labor relations, lowcountry fertilizer firms faced various legal challenges throughout the 1870s and 1880s, most of which were benign. Etiwan and Pacific Guano, however, became embroiled in lengthy lawsuits, one of which had catastrophic consequences. The Etiwan cases involved environmental pollution, and the various arguments employed revealed the pro-industrial predispositions of city and state leaders. Pacific Guano's struggles paralleled the phosphate war in the river-mining industry. Both cases illustrate the litigious nature of the three phosphate-based industries, the growing importance of lawyers therein, the continuing growth of the regulatory state, and the dangers of convergence.

While bigger was better for SASP since its creation, it was not for neighboring farmers on the Neck. In 1873, truck farmer John Kennerty complained to the South Carolina Agricultural Society and filed a lawsuit the following year arguing that fumes from SASP's sulfuric-acid works were injuring his crops. Memminger rallied to SASP's defense in the *Rural Carolinian*. While admitting that "years ago" large amounts of acid had escaped, he concluded that "now" there could be "no damage" as long as the chambers were "properly managed." In 1876, SASP settled with the farmer for $2,429 and promised to raise the exhaust pipes. Unfortunately, this did not remedy the problem, and in 1879, Kennerty appealed again to the society. Aware of the potential implications for other local fertilizer manufacturers, Dr. A. B. Rose and two others from the society seemed predisposed to favor the companies' interests over those of farmers. Rose observed that all crops seemed "in very good condition" and argued that "the mills have done all they reasonably can do." The committee concluded that "unnecessary interference" by the courts or legislature would be "a serious evil and drawback to the trade and prosperity of Charleston."[58]

Back in court in 1881, Kennerty redefined the complaint as a general environmental hazard (gases, fumes) and a public nuisance (smells) to his family, workers, and trees. He also countered Rose's bias by emphasizing truck farming's commercial importance. After farmers, including H. W. Kinsman, testified that the fumes were "very strong and offensive, affecting the eyes, nose and throat," Kennerty won a restraining order against Etiwan. The suit had become much more than a legal irritation to the company, and its legal team responded by denying environmental claims and portraying Kennerty as greedy. Admitting that the smells were "not agreeable," the lawyers argued that Etiwan's choice of the thinly populated Neck proved that it was a good corporate citizen. Kennerty's lawyer shot back, arguing that the "large and powerful corporation" was a greedy perjurer and bad neighbor and had bullied Kennerty into the earlier agreement. Most damaging, farmers living near the works had presented "frequent and strong complaint" about crop damage, bodily suffering, and "injury to their rights and property." Traditionally, Kennerty's agrarian argument—the common farmer against big business—would gain sympathy in the South, but, in what was a sign of things to come on the Neck, the judge dismissed the case. Warts and all, corporations were to remain permanent residents on, and thereby continue to transform, the Neck. While truck farmers stubbornly cultivated on the Neck for several more decades, the industrial complexes represented Charleston's smokestack future, one that would be dominated by the fertilizer industry.[59]

Pacific Guano received similar complaints about its Massachusetts plant, but the firm had more to fear from South Carolina regulators. The northern company's extensive mining of Chisolm's Island's tidal marshes led state inspectors and politicians to suspect that it was river mining. At stake was the one-dollar-per-ton royalty on river rock that Pacific Guano had so far avoided. New laws and political ambition led Comptroller General and gubernatorial candidate Johnson Hagood to pursue Pacific Guano for trespassing in 1879. Later, the Agriculture Board (under now-Governor Hagood's direction) hired Augustine T. Smythe, who possessed, they felt, "great familiarity with" the island and a thorough knowledge of "the whole subject of phosphate law."[60] Smythe and other such lawyers were becoming important players in the oft-litigious river-mining industry, but as a Coosaw Mining Company partner and lawyer, Smythe was on ethically murky ground in prosecuting a Coosaw competitor and neighbor on Chisolm's

Figure 14. Dockside rock processing, Pacific Guano, Chisolm's Island, circa 1889–95. Courtesy of Beaufort District Collection, Beaufort County Library, Beaufort, South Carolina.

Island. As the state's lawyer, he was fighting against a dangerous precedent and recovering royalties. If Pacific Guano mined its marshes royalty free, others might similarly evade the royalty in the marshy lowcountry. The state sought damages of three hundred thousand dollars, almost twice what all other companies paid in 1882. Regulators also brought suit against small producers William B. Davis and C. C. Pinckney Jr. for mining nearby Morgan Island's marshes, but Pacific Guano remained the state's main target. State officials perceived it to be a "wealthy corporation" able to spend large sums on litigation that "crippled" the budget of the state's Agricultural Department.[61]

Pacific Guano appeared to be prospering. In 1883, experts believed the firm to be worth one million dollars. No other creditors besides the state pursued the company, and its mine and factory seemed to flourish. The legal battle with the state, however, took its toll. Pacific Guano lost in circuit court in 1884 and appealed to the state Supreme Court. Litigation continued throughout the 1880s. Finally, the Agricultural Department announced in 1889 "a great triumph," a circuit court ruling for $76,874 against the company. However, it soon became a hollow victory when, "to the great surprise of the whole commercial community," the firm declared insolvency. Having helped drive one of its largest miners and manufacturers out of business, the state eventually gained only $31,250. Pacific Guano disappeared from Charleston and Beaufort Counties, with its assets sold piecemeal to satisfy creditors.[62] The company's misfortunes were unique, but they also showed the dangers of convergence; having supplied the profitable factory for decades, Pacific Guano's mines ended up killing the entire company.

Impact

The convergence of river- and land-phosphate rock mining and commercial fertilizer manufacture in the lowcountry had an enormous impact on southern farmers, the region's companies, and America's fertilizer industry. Fertilizer production soared and prices fell. Skeptical consumers as late as the 1860s, southern farmers greatly increased fertilizer consumption after 1870. While South Carolina's three industries were not the only catalysts for the increase, the uniting of mine and factory made fertilizer cheaper and more available for southern farmers. Fertilizer companies throughout the South benefited from their proximity to South Carolina's phosphate rock mines. The state's companies consumed a yearly average of twenty thousand tons of phosphate rock during the 1870s, increasing that total to one hundred thousand tons during the next decade. Charleston's companies powered South Carolina's rise from a superphosphate nonproducer in 1866 to the nation's second-largest producing state by 1883. Fertilizer shipments from Charleston grew over sevenfold between 1871 and 1884, and the rise in Beaufort was almost as dramatic. Companies in Savannah and other southern cities similarly boosted their production with South Carolina rock.[63]

The impact on the United States' fertilizer industry was just as remarkable. The phosphate rock deposits ended America's reliance on what Richard Wines terms the "urban-rural recycling system" and rapidly replaced imported guano. Wines argues that the fertilizer industry's raw material woes were not resolved until "the real breakthrough," the introduction of South Carolina rock into commercial fertilizers. Positive reviews of fertilizers manufactured in Charleston, Baltimore, and other cities made the rock popular with most U.S. manufacturers, and by 1870, 75 percent of northern manufacturers were using the state's rock. The abundant raw material helped U.S. superphosphate production jump from 102,000 tons in 1870 to 320,000 in 1880. The phosphate discovery was the catalyst in shifting the fertilizer industry's center of gravity southward from the Boston-Baltimore corridor toward Charleston and other southern cities. Insignificant until 1867, the southern fertilizer industry was born in Charleston and based "almost entirely" on treating phosphate rock with sulfuric acid. Similarly, the companies on the Neck stimulated a national evolution to large-scale manufacturing complexes. Charleston's companies were equal to all, and better than many, of the northern manufacturers, in terms of size and technology. Indeed, Wines refers to an 1880 diagram of Pacific Guano's Rikersville plant as the "typical plant" in America.[64]

Convergence had the greatest impact on southern farmers. Lowcountry river and land mining and superphosphate manufacturing made fertilizer cost effective for farmers, especially the hesitant majority in the rural South. In Charleston, prices declined dramatically at Pacific Guano (36 percent) and Wando (27 percent) between 1868 and 1873. Plentiful rock allowed, or competition forced, northern producers to drop fertilizer prices similar percentages. Lower prices helped undermine farmer skepticism of commercial fertilizers and, together with emancipation, soil exhaustion, livestock scarcities, and falling cotton prices, led to a spike in southern fertilizer consumption. Fertilizer use in southeastern states surged during the 1870s from relative insignificance to over a third of the nation's total. Georgia's consumption more than tripled between 1875 and 1884. Fertilizer dealers penetrated the Carolinas' Piedmont and other areas, leading to changes in crop selection, production and prices, debt levels, landownership, and ideas of self-sufficiency. Just as guano had ignited northern consumption decades before, South Carolina rock and the ensuing growth of the lowcountry fertilizer industry in the 1870s led to what historian Gavin Wright terms a "fertilizer revolution" in the South.[65]

Increased fertilizer consumption in the Southeast was a sign of, and catalyst for, greater changes in southern agriculture. "Land killers" before the Civil War, farming southerners began to transform themselves from "laborlords" to "landlords" in the late 1860s. The changes included cotton monoculture and heavy fertilizer use. One of the most brutal crops on soil, cotton made southern farmers dependent on fertilizer, because it increased yields in the region's difficult soil and climate conditions. The postbellum rise of "King Cotton" was intricately related to that of commercial fertilizers. Similarly, increased attention to land values led to a railroad boom and the consequent penetration of cotton and fertilizer into the Southeast's interior.[66] The decline of Charleston-centered general agents during the 1870s was a direct result. Traveling agents from Stono and the other low-country companies rode the rails to small towns throughout the South, selling fertilizer to farmers and merchants new to the cash-crop economy.

In late 1868, an anonymous writer in the *Courier* penned an ultimately perceptive conclusion trumpeting the birth of Etiwan fertilizers. He or she noted that imminent increases in mined phosphate rock would surely lower the raw material's price on the world market, and that, combined with local production, would lower fertilizer costs for the South's planters. While lamenting that American farmers consumed a comparatively small amount of fertilizer each year, the writer speculated that if South Carolina fertilizer companies increased production, prices would fall and more southern farmers would use fertilizer. The result would be a "ready market" for the local fertilizer manufacturers. Despite its promotional tone, the conclusion proved prophetic. South Carolina's increasing phosphate rock production lowered costs for fertilizer manufacturers. Lower fertilizer prices helped southern farmers overcome their doubts and dramatically increase fertilizer consumption. As early as 1872, editor D. H. Jacques of the *Rural Carolinian* was not exaggerating when he declared "It is too late in the day to say . . . that commercial fertilizers, judiciously used, do not pay."[67] The ensuing changes transformed the nature of southern agriculture, for better or worse, over the next several decades. As was common among their peers, neither writer mentioned the freedpeople who helped transform the Neck into a modern industrial district and who fashioned working-class lives within the lowcountry's thriving fertilizer industry.

Conclusions and Epilogue

The rise of land mining, river mining, and fertilizer manufacturing in South Carolina was by no means representative of the southern experience in the twenty years after the Civil War. The marriage of science, raw materials, wealth, and entrepreneurial vigor was unique, especially so early after the war. Scientists, entrepreneurs, and workers believed the discovery of phosphate to be what Francis S. Holmes termed the "providence of God."[1] As mining and manufacturing totals increased, many observers began to realize the potential benefits for the region. Touted as a revolutionary advancement in soil husbandry and successfully overcoming significant doubts, phosphate-rock-based fertilizer seemed a panacea for the South's farmers. In a larger context, however, the rise of the three industries heralded the beginning of an economic recovery for many segments of the region's population. Sporadic and weak at times and nonexistent in many areas of the South, the recovery involved tentative steps toward industrialization that led to significant changes for South Carolina, the lowcountry, the elite, and the workers. The history of the state's new phosphate-related industries, then, illuminates alternative answers to several debates within southern historiography.

The economic recovery emerging in the lowcountry in 1867 began with scientists, planters, cotton factors, freedmen, and white workers exploiting the newly discovered mineral wealth. But this part of the story was not unique. Lumber, pine tar, and textile entrepreneurs similarly exploited natural resources throughout the postbellum South. The land- and river-mining industries were different because they preceded the cotton mills and set in motion two separate chains of reaction, one agricultural and the other industrial. With the greater availability and lower cost of phosphate fertilizer in the Southeast, commercial fertilizer use in the South soared, and the option of not using fertilizer disappeared from southern agriculture. For better or worse, fertilizer increased the ability of southern farmers to grow cotton. The impact on farmers, sharecroppers, merchants, and the American fertilizer industry was profound.

The second reaction occurred in Charleston and, to a lesser extent, Beaufort and involved the birth of industrialization in areas perceived to be more wedded to the Old South than the New. Factors, merchants, and some planters transformed themselves into industrialists and converted the Charleston Neck and parts of Beaufort into industrial districts. Former slaves transformed themselves into industrial workers capable of striking. If Charleston could industrialize, would the rest of the South be far behind? Charleston and Beaufort did not become New York or Pittsburgh, but permanent change nonetheless had begun by the mid-1880s.

South Carolina's land- and river-phosphate and fertilizer industries provided some of the first evidence of a longer-term change in agriculture and industry ambiguously labeled modernization. In this case, modernization meant adopting some of the trends already apparent in the economically more prosperous and aggressive Northeast. The South did not have to "become" or emulate the North to be "modern," but certainly its adoption of fertilizer and the beginnings of large-scale industrial manufacturing signaled a break from the Old South and a move toward a more modern South. For most South Carolinians, of course, the journey to the New South began with emancipation. And for a significant number, especially those in the lowcountry, the journey continued in the phosphate, not cotton, fields (and rivers) under the direction of those who led the Old South. Continuity and change, therefore, characterized this phase of South Carolina's industrialization and added to a growing sense that complexity and diversity characterized southern industrialization.

While South Carolina conservatives managed to adapt their antebellum ideology to postwar realities and for a time regain political hegemony, they could not control and struggled to understand their former slaves. As James L. Roark reminds us, black inferiority and white supremacy were concepts that, for planters, paled in comparison to the satisfying wholeness of slavery.[2] While most freedpeople throughout the South remained mired in agricultural work, many left the fields to seek positions within the New South as industrial workers. Before and after Redemption, black South Carolinians seized the opportunities of freedom, including jobs in the land mines, on river dredges, and in fertilizer factories controlled by local elites. Although former slaves and former masters had little personal contact within the three new industries, they jointly shaped and experienced industrialization.

Conclusions

Does this case study further illuminate or answer larger questions regarding the impact of emancipation on masters without slaves, freedpeople, and the process of southern industrialization? Although it is too limited in scope to be representative of the entire South, this work does point to important trends within the South Carolina lowcountry that if found elsewhere and analyzed may have significant implications for the study of the New South. Regarding the question of continuity and southern leadership in the postbellum South, the three industries defy simple generalizations. Many of Charleston's antebellum economic elites—including the Trenholms, Memmingers, Middletons, Adgers, Draytons, and Pinckneys—survived the war to provide most of the postbellum industrial leadership. Planters and their city partners—conservative factors, merchants, and lawyers—possessed enough land, prestige, contacts, and experience to shape the new industries during the first few decades. More powerful than the middle-class businessmen involved in land and river mining and fertilizer manufacture, local elites influenced or competed with northern rivals. The alliance with the *Courier* further aided the conservative cause, as did the political activities of elites. Hardly dead after Appomattox, planters and their allies seemed reinvigorated in the 1870s.

However, the pockets of local conservatives were not deep. Capital meant influence, and while Charleston elites provided leadership, outsiders had the cash to grow and ultimately control the industries. The city's aristocracy could provide the funds to start the companies, but they could not keep up with the pace set by a "foreign-owned" company like Charleston Mining and Manufacturing. The career of Francis S. Holmes within that company foreshadowed the downfall of local elites in the three industries. While capital scarcity did not undermine the ability of lowcountry businessmen to compete successfully and occasionally dominate the industries before 1884, the chronic problem would cause them to lose control in the 1890s. Especially in the fertilizer industry, industrialization came to mean bigger and more expensive plant. Southern industrialization provided profits and some wealth to those involved, but systemic weaknesses in the lowcountry economy prevented the concentration of capital that occurred in northern cities. A familiar story throughout the postbellum South, persistent capital shortages due to emancipation and uneven economic growth eventually undermined the ability of locals to remain in control. Other southern states may have experienced a similar path to industrialization, with regional aristocrats providing the initial leadership and northern capitalists later taking over.

In this study, the impact of emancipation on the planter elite was profound. While some retained their agricultural contacts and investments, many planters, factors, merchants, and lawyers appeared "freed" to invest in industrial enterprises. They seemed transformed, and they displayed no commercial snobbery or hesitation in industrializing Charleston. Admittedly, the three new industries—dominated by local elites, located on or near plantation lands, and linked to agriculture—did not take the planters far from their roots. Still, planters and their elite allies did not flinch when directing the physical transformation of the Neck and plantation lands along the Ashley. Lawyers became an especially important part of postwar industrialization, building influence by investing in the companies and participating in the many legal and legislative battles. During 1865 to 1884, elite entrepreneurs displayed the commercial traits—initiative, aggressiveness, innovation, and a willingness to expand plant—commonly attributed to Yankee businessmen.

Why the transformation, and did other southerners outside of the lowcountry exhibit similar tendencies? Although Robert Adger and George A. Trenholm were two examples of antebellum elites eager to diversify, the planter class as a whole seemed to be shackled to slavery until 1865. The entrepreneurs within this study had, for the most part, stuck to their core businesses before the war and had exhibited little inclination to risk much on alternative economic activities. Antebellum efforts to introduce industrialization to Charleston had repeatedly stalled. The economic desperation following emancipation certainly forced the hands of lowcountry businessmen, but their sustained involvement with the land- and river-phosphate and fertilizer industries indicates that their new interest in industrialization involved more than a little passion. They liked it, and they were good at it. Lawyers obviously followed the change in interests by their aristocratic clients, but they also seemed to lead the move toward economic diversity, industrialization, and regulation. Several of the entrepreneurs expanded industrial opportunities beyond the lowcountry and the state, making and expanding contacts with elite and middle-class businessmen elsewhere. For example, leaders of the Coosaw, Atlantic, and Oak Point companies were early investors in coal and iron companies in Alabama. The ensuing intrastate and interstate alliances suggest the existence of elite networks throughout the South, sharing conservative tenets and pooling capital and experience to create industries in the region. Such networks carried forward the urge, so obviously exhibited by Trenholm, to marry southern patriotism with profits and may have contributed to the start of the New South in certain locations as early as the 1870s.[3]

This case study provides evidence that southern industrialization was a distinctly southern process, led initially by the planter elite and strongly influenced by the black labor force. The political, economic, and social context of the nineteenth-century South contributed to the creation of a process of industrialization vastly different from the northern model. That the southern version came later, was weaker, had fewer linkages, and was less transformative than its northern cousin does not mean that it was not significant. More study is needed on pockets of industrialization throughout the South, including instances of failed and short-lived industries, in order to fully understand southern entrepreneurs, workers, and industrialization.

Within the 1865–84 period, local elites took strong action to undermine northern domination of the three industries and the creation of a colonial economy in the lowcountry. Had South Carolina merely been a rock supplier for northern and European manufacturers, others would have controlled its industrial destiny. However, the development of a viable fertilizer-manufacturing industry that included local and outside ownership undercut the still-forming colonial relationship. Far from selling out their section to "foreigners," local conservatives and their political allies consciously fought to control the land- and river-mining and fertilizer-manufacturing industries and to keep profits in South Carolina. Among their many motivations was a sense of patriotism for the South. Former Confederates now sought to empower their section economically. This part of the story is consistent with the strong leadership displayed by the planter aristocracy in all three industries. Despite capital shortages that would eventually undermine their claims to leadership and control, elites actively fought for an economic version of home rule.

Beyond analyzing aspects of entrepreneurship and industrialization, this case study exposes deep divisions within political parties and illustrates the impact of discord on public policy during Reconstruction and Redemption. The political war over river mining was especially revealing. Carpetbaggers, scalawags, and freedpeople had radically different visions for postwar South Carolina under Republican rule, and all three groups used river mining as a vehicle to advance their goals for the party. Likewise, fissures within the Democratic Party existed before and after Democrats returned to political power in 1876. The traditionally Democratic business community in Charleston seemed to be a remarkably flexible group in cooperating with the alleged enemy, northern Republicans. Rather than hiding behind a party barricade, Republicans and Democrats cooperated in political and economic self-interest. A similarly fluid situation existed regarding race, at least during Reconstruction. When business opportunities beckoned, whites, even those within the reactionary South Carolina low-country, struggled with greed for personal and sectional profit and with fear of political and racial amalgamation. Although state policies reflected these vacillations, with generally negative results for long-term economic development, the overall impact was to limit, slow, or at least delay an invasion of northern investors in the three new industries. While South Carolina's politics often defied comparison, the interplay of politics and economics suggests potentially fertile ground for study in other states.

Freedpeople also contributed to making southern industrialization a distinctly southern process, and their roles in the three industries demonstrate the multiple consequences of emancipation. This study makes a significant contribution to the study of historically "invisible" black workers on the edge of urban and rural working environments. While sharecropping and other postwar agricultural arrangements trapped many rural freedpeople and their descendents, evidence from the 1870 and 1880 censuses suggests that freedpeople created innovative economic alternatives that frustrated and confused white employers and statisticians. More investigation is needed into the importance of hunting, fishing, and other aspects of the rural black underground economy.[4] Land- and river-phosphate mining and fertilizer manufacturing provided opportunities not uniformly available throughout the South, but their existence suggests that other equally flexible options existed for freedpeople.

Black miners and factory workers appeared to have different values than those the southern elite and northern prophets of free labor promoted. Of course planters such as Williams Middleton preferred freedmen to act like slaves, but other members of the planter aristocracy merely hoped for black workers to adopt the Protestant work ethic, a hope shared by carpetbaggers and other northern commentators. Almost all white employers throughout the South were disappointed with the freedpeople's work habits, which they interpreted as "resistance," "laziness," or "disrespect." Employers within the three industries were no different, frequently complaining that black laborers appeared to be ignorant of or spurn economic rationality, self-discipline, and market incentives. Only the most perceptive white observers realized that the freedpeople were acting upon radically different priorities, including autonomy, family, and education. Lowcountry blacks wanted to "get ahead," but their path to that goal involved more than mere economic gain. To freedmen, having one full-time job made them dependent on a white employer and exposed them to control and coercion reminiscent of slavery. Making one adult male the sole breadwinner exposed the black family to the precarious winds of racism and the white economy. Recognizing that their lives had become cheap in the white mind after emancipation, black people feared that one misstep by the only wage earner could lead to a lost job, a lynching, and economic disaster for the family. Various family members participated in multiple economic activities in order to spread the risk, increase free time, decrease white control and contact, add flexibility, and improve the chances of the family's survival.

As whites throughout America became increasingly worried that the economic world of the freedpeople might endanger the social order, blacks responded by making as much of their working lives "invisible" to whites as possible. Miners opted not to inform their employers when they would show up to work, preferring to come and go at irregular intervals. When the repression of Redemption followed the uncertainties of Reconstruction, freedpeople responded in their work patterns to the continuing lack of trust many felt for southern white employers. Less evidence exists on the fertilizer laborers' work habits, but they too likely used the threat of irregular working patterns to extract concessions. Declining trust was not universal, however. In individual cases where trust improved over time, stability appears to have replaced irregularity.

The origins and actions of laborers in the three industries suggest that they developed a degree of working-class consciousness. While some may have worked at nonagricultural and urban jobs as slaves, most of the workers (or their parents) came from rural plantations, where subtle forms of resistance predominated. After emancipation and the phosphate discovery, these laborers moved in and out of Charleston but did not immediately transform themselves into a self-conscious working class. Their sense of class mirrored the environment. Before fertilizer companies settled on the Neck, Charleston's business district was more commercial than industrial. As the industry grew and black fertilizer workers spent more time in the city, they were, as the 1873 strike illustrated, beginning to perceive themselves as industrial workers. That year they joined others from the city's emergent industrial district to demand, not subtly maneuver for, better wages. But the Neck was close enough to rural plantations that elements of the flexible economic life of the land miners may have influenced the actions of the fertilizer workers.

In a more subtle way, land- and river-phosphate miners absorbed elements of working-class consciousness. Most miners seemed to have adopted a variety of subsistence, agricultural, wage, and bartering enterprises with which to live, but some may have worked at industrial jobs during sojourns into Charleston or Beaufort. While they preferred the "weapons of the weak" to formal strikes and unions, their obstruction of white-initiated change, especially with regard to labor systems and autonomy, seemed more aggressive than what had occurred during slavery. Company housing was a necessary but often temporary concession for land miners in 1870, but by 1880, they appeared to have shunned it. In short, lowcountry miners merged their rural and urban experiences in their struggle to create as separate a working world as possible and to keep their personal lives as private as they could.

Evidence from entrepreneurs, managers, and workers in the three industries points toward some general conclusions regarding the New South. The Civil War and especially emancipation destroyed the Old South and made a large impact on the South's economy. Overturning slavery as the South's labor system, emancipation brought a host of immediate and long-term changes to elites and their former slaves. Planters lost a significant percentage of their wealth, as well as a more controlled and naturally reproducing labor force. Freedpeople gained mobility and a host of partial freedoms. Social, political, religious, and economic changes would follow in the coming decades, including new industries and new ways of organizing agricultural labor, but 1865 did not mark a clean slate. A strong sense of continuity remained. Black people predominated as the working class, and white elites retained control of the land and the business world. A New South began forming almost immediately after the war that carried forward the burdens and actors of the past even while it forced most southerners to look ahead hopefully to a safe and prosperous future. The strange combination of continuity and change shaped the beginnings of southern industrialization and its participants.

Besides debating the levels and implications of continuity and change, historians have dubbed various eras "the New South." A popular choice for the "first" New South remains the period 1877–1913, bookended by Redemption and World War One. Howard Rabinowitz describes the years 1865 to 1876 as the period when the "roots" of the New South began to form.[5] This study reinforces the need to examine more closely the first decade following the war as an important economic time for the South, a time when surviving antebellum elites sought to reassert economic leadership. Likewise, freedpeople took advantage of the many opportunities during Reconstruction (and those opportunities that survived Redemption) to assert their sense of freedom, independence, and mobility. The immediate postwar period featured many struggles between former masters and slaves but also many instances of cooperation. Carpetbaggers, scalawags, mulatto elites, and others contributed to the primary battle waged between elites and freedpeople. A strong emphasis in many histories on the politics of Reconstruction and Redemption tends to obscure the importance of economic struggle. The decades after the war are crucial in understanding all future New Souths, and the early industries, failed and successful, are crucial in understanding the course of southern industrialization.

Epilogue

Previous historians have underestimated the impact of the three industries. Don H. Doyle and Tom W. Shick cite the events of the 1890s, including the drastic declines in land and river mining and the takeovers of local fertilizer companies by national trusts, in arguing that the prosperity so apparent by 1884 failed to lead to "sustained economic development" in the lowcountry. They assert that "obstructions" by black laborers and white elites were two of the reasons that the once-promising industries degenerated into "a harmless flurry." Blacks workers' pursuit of an "independent course" in agriculture and other occupations hurt the industries' development. And the city's old families, with their "fiercely conservative resistance to things new," participated in the industries more out of financial "desperation" than enthusiasm for New South values. Shick and Doyle lament that the "boom" failed to transform the elite into "a vigorous bourgeoisie imbued with a spirit of bold entrepreneurship," and Doyle expounds on this "Charleston Style" in a subsequent work.[6]

A closer look at the industries and participants reveals many of these conclusions to be flawed. Only Williams Middleton could claim that the freedpeople created permanent obstructions, and even he embraced, albeit unsuccessfully, Charleston's industrial future. While miners and mill workers maneuvered for housing and other concessions from management, they made the primary contributions to the rapidly increasing production totals. They struck and walked away, but there was no labor shortage. Charleston's elites, young and old, appeared motivated by the desire to make money and to grow the three industries. Well-established antebellum leaders, such as C. G. Memminger, George A. Trenholm, St. Julien Ravenel, and Robert Adger, provided energetic and effective postwar guidance, and many others made themselves into industrial entrepreneurs. Atlanta may have been the more dynamic New South city, but Charleston's leaders and laborers did not shy away from the challenges.

The three industries led to sustained economic development. Phosphate land mining and, to a lesser extent, river mining were the catalysts for development of fertilizer companies on the Charleston Neck, and those factories were the first signs of lowcountry industrialization—large industrial complexes, railroad links, factory neighborhoods, additional industries, and strikes. In 1906, black and white fertilizer workers struck for better wages. Charleston led the world in fertilizer production for several years before and after World War One and remained a major fertilizer producer into the 1960s. A United States Navy base and other factories joined the fertilizer companies in shaping the industrial landscape on the Neck and Charleston's docks throughout the twentieth century. During the 1960s, United States Congressman Mendel Rivers greatly expanded existing military facilities near the industrialized Neck, and companies such as Dupont, Lockheed, and General Electric shared the roads, rails, and rivers with fertilizer factories. Housing for industrial and defense workers grew proportionately on the Neck. In sum, the phosphate discovery of 1867 led to more than a century of lowcountry economic development, not just a "boom."[7]

River and land mining died out in the lowcountry due to a combination of bad politics, bad luck, and bad weather, not the actions of local elites and workers. Political grandstanding by Governor Benjamin R. Tillman exacerbated problems in the early 1890s stemming from the discovery of Florida's larger reserves and from a severe hurricane. In his drive to control the state, "Pitchfork Ben" targeted conservatives through regulation of the river-mining industry. Tillman and the Coosaw Company became entangled in a court battle that forced Coosaw to suspend its mining operation in March 1891. In the thirteen-month interval, European fertilizer manufacturers previously loyal to "Carolina river rock" switched to Florida rock. When Coosaw and the state ended their phosphate war, Florida's higher-quality and more abundant rock had won many admirers. In August 1893, a hurricane centered in the river-mining district destroyed plant, killed miners, and ended mining companies' efforts to recover market share from Florida. The Florida effect and diminishing reserves also crippled the South Carolina land-mining industry before 1920. Although river and land mining continued until the 1937–38 season, South Carolina lost its place as the world's preeminent phosphate source before 1900. The days of phosphate mining in the state had been numbered after the Florida discovery, but Tillman's attack, which even he later regretted, hastened its demise.[8]

Figure 15. Stono fertilizer factory, Charleston, circa 1890. Includes boat and rail transport, managers and workers, rock and storage shed, main building and sulfuric acid facility. By permission of South Caroliniana Library, University of South Carolina, Columbia.

Fertilizer plants maintained their place at the core of the industrialized Charleston Neck for a century, but Charlestonians lost control of the companies before the dawn of the twentieth century. During the Great Merger Movement of the 1890s, Charleston's fertilizer entrepreneurs did not have the money to control their companies' destinies. The Virginia-Carolina Chemical Company bought most of Charleston's fertilizer and land-mining companies and relegated local conservatives to positions as branch managers or salesmen. Capital scarcity had finally caught up with Charleston's elite. The companies remained in the area, but control and profits left town.[9]

In the 1980s and 1990s, the Neck and adjacent areas began a slow process of deindustrialization, and the U.S. Environmental Protection Agency moved in. EPA officials named the Neck "one of the most concentrated areas of contamination" in America and targeted several of the fertilizer factory sites for Superfund status. Symbolic of the environmental problems

stemming from the fertilizer era, a shrimp caught in the Ashley River near the Neck exploded. The crustacean had become coated with phosphorus leaching from abandoned fertilizer factories and had, when it dried on the shore, spontaneously caught fire. Among the sites targeted for cleanup are those once occupied by the Wando, Stono, Atlantic, Ashepoo, Pacific Guano, and Etiwan companies.[10]

Having created sustained industrial development, the river-mining, land-mining, and fertilizer-manufacturing industries left many legacies. While the pollution was a grave legacy, the magnitude of the problem indicates the scope, duration, and importance of the fertilizer industry's decades on the Neck. It provided jobs and lured capital into the lowcountry for over a century and helped attract similar industries to the area. A second and equally troublesome legacy, the sheer tonnage of fertilizer produced in the area altered agricultural production patterns, especially in the cotton-obsessed South. By providing the means for southerners to overproduce the crop, fertilizer contributed to a century of agricultural poverty. A final and ambiguous legacy involves the lowcountry's black working class. Freedpeople and their descendants used the three industries as part of life strategies geared toward autonomy. Perhaps these industrial jobs shielded them from more exploitative occupations, but in the long run mining and manufacturing did not noticeably improve their economic standing.

Southern industrialization was indeed a mixed blessing, but it included the active participation and collaboration of elites and laborers and constituted a major change for the region. Although minor players in the drama of the postwar South, the phosphate land- and river-mining and fertilizer-manufacturing industries continue to have an impact on the low-country, and they provide insight into areas and issues previously overlooked by historians. Far from being a boom in the stillbirth of the New South, South Carolina phosphate was the midwife to Charleston's peculiar industrialization.

Abbreviations

Acts	South Carolina General Assembly, *Acts and Joint Resolutions of the General Assembly of the State of South Carolina*, Columbia
BTL	Baker and Turner Letterbook, Middleton Place Archives, Charleston
CDC	*Charleston Daily Courier*
Census 1870	U.S. Bureau of the Census, *Ninth Census of the United States, 1870: Manuscript Schedule*
Census 1880	U.S. Bureau of the Census, *Tenth Census of the United States, 1880: Manuscript Schedule*
CM	*Charleston Mercury*
CMMC	Charleston, South Carolina, Mining and Manufacturing Company
CNC	*Charleston News and Courier*
CPC	*Charleston Post and Courier*
EWS	Edward Willis Scrapbooks, Charleston Museum Archives, Charleston
JDP	John Drayton Papers, South Carolina Historical Society, Charleston
JSH	*Journal of Southern History*
MP	Middleton Place Archives, Charleston
MSP	Mitchell & Smith Papers, South Carolina Historical Society, Charleston

Pratt Papers	Nathaniel A. Pratt Papers, Perkins Library Special Collections, Duke University, Durham, N.C.
RC	*Rural Carolinian*
Reports	*Reports and Resolutions of the General Assembly of the State of South Carolina*, Columbia
RTS	Robert Turner & Son
SASP	Sulphuric Acid and Super-Phosphate Company
SCHM	*South Carolina Historical Magazine*
SCHGM	*South Carolina Historical and Genealogical Magazine*
SCHS	South Carolina Historical Society
SCIPL	*South Carolina Institute Premium List, Fair of 1870*, Charleston: Walker, Evans & Cogswell, 1870
UNC-CH	Southern Historical Collection, University of North Carolina–Chapel Hill, Chapel Hill, N.C.
WM	Williams Middleton
WML	Williams Middleton Letterbook 1868–70, transcribed by Barbara Doyle, Middleton Place Archives, Charleston
WMP	Williams Middleton Papers, Middleton Place Archives, Charleston

Notes

Introduction

1. Coclanis, *Shadow*, 106, 112, 115, 117–18, 125.

2. Fraser, *Charleston!*, 244.

3. Willis, "Marl Beds," 47; "The Charleston Phosphates," *SCIPL*, 68, EWS-3; Genovese, *Two Essays*, 123; Genovese, *Political Economy*, 13–14, 19–20, 23–25, 28–31, 180–88, 191–201; Oakes, *Ruling Race*, xi–xiii; Cobb, "Beyond Planters and Industrialists," 48, 51; Kaye, "Second Slavery," 628–29; Prymak, review, 97–99; Gagnon, review.

4. Bateman and Weiss, *Deplorable Scarcity*, 161–63; Genovese, *Political Economy*, 28–31.

5. Cooper, *Conservative Regime*, 17–20, 39–40, 132–33.

6. Holden, *Maelstrom*, 2, 6, 111–14; Holden, "'Public Business,'" 124–26; D. Doyle, *New Men*, 117. Don H. Doyle argues that Charleston's elites used their social status to fend off much of what the New South brought to more progressive southern cities.

7. Wiener, *Social*, 5, 9, 199–203; Mandle, *Roots*, 21–31; Cobb, "Beyond Planters," 46–48; Woodward, *Origins*, 19–22, 107–8, 148–53, 291–92, 309–10; Hackney, "Retrospect," 193–96; Rabinowitz, *First New South*, 18–19. Continuity proponents Wiener and Mandle argue that planters dominated the southern economy after the war. Change advocate Woodward sees "new men" from the middle class emerging during Redemption to lead the South's business community. Woodward also argues that capital-poor southern industrialists were vulnerable to northeastern businessmen who plundered the South's raw materials, manufactured the refined products elsewhere, and kept the profits for themselves.

8. Rabinowitz, *First New South*, 5–71; Carlton, *Mill and Town*, 4–10; Flamming, *Creating the Modern South*, xxi–xxv; G. Wright, *Old South, New South*, 156–65.

9. Mandle, "Black Economic Entrapment," 69–75. Mandle argues that the vast majority of freedpeople were trapped in an exploitative agricultural labor system after the Civil War.

10. Foner, *Reconstruction*, 55–56, 132; Kelley, "'We Are Not What We Seem,'" 76–80; Arnesen, "Up from Exclusion," 146–47, 156–62.

11. Foner, *Reconstruction*, 77; Stanley, *From Bondage to Contract*, 62–84, 138–39, 175; H. Richardson, *The Death of Reconstruction*, ix–xv, 99–101.

12. Wetherington, *Wiregrass Georgia*, xviii–xxii, 99–107, 117–22, 162–65, 240–41; Powers, *Black Charlestonians*, 1–8, 105, 126, 133, 168–69, 250; Saville, *Work of Reconstruction*, 1–4, 42, 111–13; Rabinowitz, *First New South*, 42–43. Reflecting census weaknesses, Powers and Saville largely ignore the phosphate and fertilizer industries.

13. Coclanis, *Shadow*, 128, 139.

14. Shick and Doyle, "Boom," 1–4.

Chapter 1. Antecedents, Precedents, and Continuities, 1800–1865

1. F. Holmes, *Phosphate Rocks*, 7, 27–35; Sanders, "Additions," 10–12; Sanders, unpublished diagram, Charleston Museum, 1999. Before "phosphate rock" became the standard term around 1870, commentators used many other names to describe the mineral, including marl-rock, marl-stone, bone-phosphate, coprolites, conglomerates, and bone-rocks.

2. Willis, "Marl Beds," 47, 80; F. Holmes, *Phosphate Rocks*, 3, 26, 31, 70; Pratt, *Ashley River Phosphates*, 42. Holmes estimated the Ashley River phosphate stratum to average fifteen to eighteen inches deep and yield six hundred tons per acre.

3. Wines, *Fertilizer in America*, 162–63.

4. Wines, *Fertilizer in America*, 3–5, 33–36, 162–69; W. Jordan, "Peruvian Guano Gospel," 211.

5. Wines, *Fertilizer in America*, 3–4, 157–59, 165–66.

6. Wines, *Fertilizer in America*, 33–47, 162–65.

7. Wines, *Fertilizer in America*, 45–46; R. Taylor, "South Carolina," 179–89.

8. Wines, *Fertilizer in America*, 24, 83–87, 108–11, 165–70; R. Taylor, "Southeast Part I," 308; Blakey, *Florida Phosphate*, 1–3; F. Holmes, *Phosphate Rocks*, 51. After 1867, marketers used the term "superphosphate" to describe phosphate rock dissolved in sulfuric acid. Roughly equal, modern and nineteenth-century BPL numbers measure the amount of phosphorus available to plants in the fertilizer.

9. Wines, *Fertilizer in America*, 70–75, 124–26, 163–70; Blakey, *Florida Phosphate*, 6–7.

10. W. Jordan, "Peruvian Guano Gospel," 211–221; Wines, *Fertilizer in America*, 41–42, 157–59; Genovese, *Political Economy*, 93.

11. W. Jordan, "Peruvian Guano Gospel," 211–221; Wines, *Fertilizer in America*, 41–42, 157–59; G. Wright, *Old South, New South*, 17–34.

12. Genovese, *Political Economy*, 26–27, 85–90.

13. G. Wright, *Old South, New South*, 17–34; Genovese, *Political Economy*, 95.

14. Genovese, *Political Economy*, 90–99; Faust, *James Henry Hammond*, 114–15, 128, 236; Allmendinger, *Ruffin: Family and Reform*, 31–34, 73–74, 114–15, 118–28; Egerton, "Markets Without a Market Revolution," 207–21; Allmendinger, "The Early Career of Edmund Ruffin," 127–54.

15. R. Taylor, "South Carolina," 179–89; R. Taylor, "Southeast Part I," 305, 309–11;

R. Taylor, "Sale and Application," 46; Genovese, *Political Economy*, 94–95; McPherson, *Battle Cry of Freedom*, 99–102.

16. R. Taylor, "South Carolina," 179–89; Genovese, *Political Economy*, 94–95.

17. R. Taylor, "South Carolina," 184–85; Wines, *Fertilizer in America*, 41–42; Shepard Jr., "Report 1870," 3–10.

18. R. Taylor, "South Carolina," 184–85.

19. Stephens, *Ancient Animals*, ix, 1–2; D. Taylor, *Naturalists*, 2–9; Sanders, "Additions," 10–12, 17–18, 25–26; Sanders, unpublished diagram.

20. Stephens, *Ancient Animals*, ix, 1–2; D. Taylor, *Naturalists*, 2–9, 222.

21. Stephens, *Ancient Animals*, 3–5, 44; "Early Notice of the Charleston Phosphates," *RC* 1, no. 10 (July 1870): 640–41; Willis, "Marl Beds," 46.

22. Stephens, *Ancient Animals*, 3, 44; F. Holmes, *Phosphate Rocks*, dedication page, 10–12, 56–57; P.H.M., "Earliest Notice of the Phosphates," *RC* 1, no. 11 (August 1870): 699–700.

23. D. Taylor, *Naturalists*, 122–23; Wines, *Fertilizer in America*, 19; Allmendinger, *Incidents*, 1, 64, 67; B. Mitchell, *Ruffin*, 13–14, 45–48; Allmendinger, *Ruffin: Family and Reform*, 29; Blakey, *Florida Phosphate*, x; Mathew, *Crisis*, 226. Marl (or "calcareous manure") contained fossil shells and soil and combined clay and carbonate of lime. By decreasing soil acidity, it allowed the soil to absorb, rather than leach, added nutrients.

24. Ruffin, appendix 50–52; F. Holmes, *Phosphate Rocks*, 56–57, 65–66; Mathew, *Crisis*, 39–40, 88; Allmendinger, *Incidents*, 67–69, 176–77; Stephens, *Ancient Animals*, 3–5; B. Mitchell, *Ruffin*, 1, 48; Willis, "Marl Beds," 47.

25. Stephens, *Ancient Animals*, 3–5, 9; Mathew, *Crisis*, 129–39; "Early Notice," *RC* (July 1870): 640–41; Francis S. Holmes, "Calcinated Marl as a Fertilizer," *RC* 5, no. 8 (May 1874): 402–3; F. Holmes, *Phosphate Rocks*, 45.

26. F. Holmes, *Phosphate Rocks*, 46.

27. Mathew, *Crisis*, 36; Stephens, *Ancient Animals*, 5; D. Taylor, *Naturalists*, 144–45; H. Johnson, "Background," 5–6; Tuomey, *Survey* (1844), 53–55; Tuomey, *Report* (1848), iii.

28. Tuomey, *Survey* (1844), 53–55; F. Holmes, *Phosphate Rocks*, 7–9, 58, 63; "Charleston Phosphates," *SCIPL*, 69.

29. Tuomey, *Survey* (1844), 53–55.

30. Tuomey, *Report* (1848), 164–65; D. Taylor, *Naturalists*, 144–45; Willis, "Marl Beds," 47–48.

31. F. Holmes, *Phosphate Rocks*, 47.

32. Tuomey, *Report* (1848), 235, appendix xxxiv–xxxviii; Mathew, *Agriculture*, 337–38; Robinson, "Charles Upham Shepard," 85–103; *Year Book—1886, n.p.*; Charles U. Shepard Jr., "Soils," *RC* 3, no. 12 (September 1872): 646.

33. Tuomey, *Report* (1848), 235, appendix xxxiv–xxxviii; Stephens, *Ancient Animals*, 4.

34. Tuomey, *Report* (1848), 235, appendix xxxiv–xxxviii; Stephens, *Ancient*

Animals, 4; Blakey, *Florida Phosphate*, 6–7, 11; Wines, *Fertilizer in America*, 72; F. Holmes, *Phosphate Rocks*, 57; "Charleston Phosphates," *SCIPL*, 69.

35. Tuomey, *Report* (1848), 235, appendix xxxvii; Pratt, *Ashley River Phosphates*, 15; F. Holmes, *Phosphate Rocks*, 64–65; Marcus, "Setting the Standard," 48–61.

36. Charles U. Shepard Sr., "Notice of the Guanape Guano," *RC* 1, no. 8 (May 1870): 469; "Charleston Phosphates," *SCIPL*, 70; F. P. Porcher, "Popular View of South Carolina Phosphates," *CDC*, 19 August 1870, 2.

37. "Charleston Phosphates," *SCIPL*, 70; Charles U. Shepard Sr., "The Charleston Phosphates," *Address before Medical Association of the State of South Carolina* (1859), 3–4, Wando Mining and Manufacturing Company pamphlet, EWS-1; Chazal, *Century*, 38–39; Simkins and Woody, *South Carolina During Reconstruction*, 305; Willis, "Marl Beds," 50.

38. Lewis M. Hatch, "A Contribution to the History of the Charleston Phosphates," *RC* 2, no. 6 (March 1871): 357–58; Willis, "Marl Beds," 50; Chazal, *Century*, 39.

39. Hatch, "Contribution," 357–58; Chazal, *Century*, 38–43; Charles U. Shepard Sr., letter 7 November 1868, Wando Mining and Manufacturing Company pamphlet, EWS-1; Willis, "Marl Beds," 50.

40. George T. Jackson (Augusta) to Charles U. Shepard Jr. (Charleston), 11 July 1873, in Chazal, *Century*, 41; Chazal, *Century*, 39, 41–43; Willis, "Marl Beds," 49–50; Hatch, "Contribution," 358.

41. Donnelly, "Scientists," 70, 78.

42. Stephens, *Ancient Animals*, 22–28, 44–45; Wines, *Fertilizer in America*, 116–19; Donnelly, "Scientists," 78–79.

43. "Dr. N. A. Pratt, Scientist and Builder," 55–56; Wines, *Fertilizer in America*, 116; Donnelly, "Scientists," 70, 76–79; Schroeder, "'We Will Support,'" 302–4; *City of North Charleston Historical and Architectural Survey*, 60–61.

44. Stephens, *Ancient Animals*, 22–28, 44–45; Wines, *Fertilizer in America*, 116–19; F. Holmes (Charleston) to Pratt (Charleston), 17 September 1868, Pratt Papers; Pratt to F. Holmes, 18 September 1868, Pratt Papers; "Charleston Phosphates," *SCIPL*, 72; Pratt, *Ashley River Phosphates*, 12–13; F. Holmes, *Phosphate Rocks*, 63–69.

45. "Dr. St. Julien Ravenel," *CNC*, 17 March 1882, EWS-6; E. B. Richardson, "Dr. Anthony Cordes" (1942), 228–29; E. B. Richardson, "Dr. Anthony Cordes" (1943), 32–33; Holman, "Summer of 1841," 1–3; Davidson, *Last Foray*, 3–4; Sanders and Anderson, *Natural History Investigations*, 66–68, 70, 72; Poston, *Buildings of Charleston*, 218, 268; Stephens, *Ancient Animals*, 47–48.

46. "Dr. St. Julien Ravenel," *CNC*, 17 March 1882, EWS-6; Holman, "Summer of 1841," 2; Mathew, *Crisis*, 116.

47. "David C. Ebaugh," 32–36; Coker, *Maritime*, 256–62; Sass, "The Story of Little David," 620–25; Hamrick, "To Sink a Yankee Ship," 23–30; Poston, *Buildings of Charleston*, 56, 409.

48. Capers, *Memminger*, 7–22, 24–35, 201–2, 242–43, 286–91, 345–65, 369, 382–84; M. Johnson and Roark, *No Chariot Let Down*, 40–43.

49. Capers, *Memminger*, 369; Woodman, *King Cotton and His Retainers*, xii, 4; Coker, *Maritime*, 197.

50. H. Holmes, "The Trenholm Family," 153–55; Coker, *Maritime*, 198–200; Nepveux, *Trenholm*, 6–7; Wise, *Lifeline*, 9, 46, 334; Loy, "10 Rumford Place," 349, 355–56.

51. Coker, *Maritime*, 198–200; E. B. Richardson, "Dr. Anthony Cordes" (1942), 228–29; E. B. Richardson, "Dr. Anthony Cordes" (1943), 32–33.

52. Davidson, *Last Foray*, 3–4; M. Johnson, "Planters and Patriarchy," 62–64; Poston, *Buildings of Charleston*, 53–54; newspaper clippings and obituaries, Adger Family Papers, (within Smyth Stoney Adger Collection), SCHS; Stevenson, *Diary of Clarissa Adger Bowen*, 57.

53. Poston, *Buildings of Charleston*, 53–54; Fraser, *Charleston!*, 261–62; Coker, *Maritime*, 187–90, 200, 268, 278–83; Wise, *Lifeline*, 69, 114, 221.

54. Poston, *Buildings of Charleston*, 89, 581–82; Coker, *Maritime*, 271–78; Bulloch, *Secret Service*, 52–53, 70–71; Spence, *Treasures of the Confederate Coast*, 9–29; Ethel Trenholm Seabrook Nepveux, e-mail to author, 6 February 2003; M. Mitchell, *Gone With The Wind*; Ripley, *Scarlett*, 210, 230, 241–42. Spence and Nepveux claim that Mitchell based Rhett Butler on Trenholm. In Ripley's sequel, Butler is a land-phosphate mining entrepreneur on an Ashley River plantation.

55. Coker, *Maritime*, 289–92; Loy, "10 Rumford Place," 350–54, 360–61, 363–66, 371–74.

Chapter 2. The Creation of Industry and Hope, 1865–1870

1. Stephens, *Ancient Animals*, 28–31; Warren, *Joseph Leidy*, 139.

2. Loy, "10 Rumford Place," 369–74.

3. Coclanis, *Shadow*, 111–12, 128–29; Fraser, *Charleston!*, 268.

4. Willis, "Marl Beds," 47; Pratt, *Ashley River Phosphates*, 42.

5. Wines, *Fertilizer in America*, 112–14.

6. Lewis M. Hatch, "A Contribution to the History of the Charleston Phosphates," *RC* 2 (March 1871): 358; Robinson, "Charles Upham Shepard," 85–103; *Year Book—1886*, 218.

7. "Dr. St. Julien Ravenel," *CNC*, 17 March 1882, EWS-6; Willis, "Marl Beds," 50–51; "The Charleston Phosphates," *RC* 1 (October 1869): 48; "David C. Ebaugh," 32–36; Wines, *Fertilizer in America*, 115–16. Rarely specified in contemporary sources, a company's "capital" described paid-in capital, pledged capital, or potential capital.

8. Chazal, *Century*, 42–44; Willis, "Marl Beds," 50–51; Hatch, "Contribution," 357–58; Charles U. Shepard Sr. (Charleston), letter 7 November 1868, Wando Mining and Manufacturing Company pamphlet, EWS-1; "Bird's Eye View" map; Wines, *Fertilizer in America*, 116; Stephens, *Ancient Animals*, 44; Pratt, "Present," 227; "Scientists Rekindle Dispute Over Island," *Charlotte Observer*, 11 September 1998, 10A.

9. Pratt, *Ashley River Phosphates*, 12–13; Blakey, *Florida Phosphate*, 12; Pratt, "Present," 227.

10. Pratt, *Ashley River Phosphates*, 13–14; Pratt (Charleston) to F. Holmes (Charleston), 18 September 1868, Pratt Papers; Pratt, "Present," 227–28; "Charleston Phosphates," *SCIPL*, 72–74; Blakey, *Florida Phosphate*, 12; Wines, *Fertilizer in America*, 116; Chazal, *Century*, 44.

11. Willis, "Marl Beds," 50–52; Waggaman, "Report," 2; Jakes, *Heaven and Hell*, 410–18; Shaler, *Phosphate Beds*, 16; Pratt, "Present," 228–29.

12. Pratt, "Present," 227. For Pratt's methods of chemical analysis, see Pratt, *Ashley River Phosphates*, 22–23.

13. Pratt, *Ashley River Phosphates*, 14–15; Willis, "Marl Beds," 55; F. Holmes, *Phosphate Rocks*, 64–66; F. Holmes (Charleston) to Pratt (Charleston), 17 September 1868, Pratt Papers; Pratt (Charleston) to F. Holmes (Charleston), 18 September 1868, Pratt Papers.

14. Blakey, *Florida Phosphate*, 2, 146; F. Holmes, *Phosphate Rocks*, 67–68; Pratt, *Ashley River Phosphates*, 13, 15; F. Holmes to Pratt, 17 September 1868, Pratt Papers.

15. F. Holmes, *Phosphate Rocks*, 67–68; Pratt, *Ashley River Phosphates*, 15–16; F. Holmes to Pratt, 17 September 1868, and Pratt to F. Holmes, 18 September 1868, Pratt Papers; "The Charleston Phosphates," *SCIPL*, 71, EWS-3; "Charleston Phosphates," *RC* (1869): 46; Pratt, "Past," 149, and "Present," [224?].

16. Pratt, "Present," 227–28.

17. F. Holmes, *Phosphate Rocks*, 68–70; "The Second in a Series of Lectures," *CDC*, 28–29 August 1867, 2; Pratt, *Ashley River Phosphates*, 15–17; F. Holmes to Pratt, 17 September 1868, and Pratt to F. Holmes, 18 September 1868, Pratt Papers; Pratt, "Present," 228.

18. Poston, *Buildings of Charleston*, 21, 28; Fraser, *Charleston!*, 195, 262–69, 275.

19. Edgar, *South Carolina*, 285; Coclanis, *Shadow*, 126–30, 154–55; Fraser, *Charleston!*, 196.

20. Pratt, "Present," 227–28.

21. Pratt, "Present," 227.

22. F. Holmes, *Phosphate Rocks*, 69; F. Holmes to Pratt, 17 September 1868, and Pratt to F. Holmes, 18 September 1868, Pratt Papers; Painter, "Recovery of Confederate Property," 139, 422–23, 425, 436, 459; Wise, *Lifeline*, 222–24; Bulloch, *Secret Service*, 423–27; Nepveux, *Trenholm*, 96–98. Trenholm and his associates remained under investigation for their Confederate activities. Welsman may have been a conduit to Holmes from Trenholm.

23. Foner, *Reconstruction*, 276–80.

24. Willis, "Marl Beds," 51–52.

25. "Charleston Phosphates," *SCIPL*, 75.

26. F. Holmes to Pratt, 17 September 1868, and Pratt to F. Holmes, 18 September 1868, Pratt Papers; Mrs. St. Julien Ravenel, *Charleston*, 320–24; Kilbride, "Southern Medical Students," 698–99, 709–16, 731.

27. Stephens, *Ancient Animals*, 5–15, 22; "Super-Phosphate of Lime," *CDC*, 27 February 1869, 4; Wines, *Fertilizer in America*, 117–18, 215n35; Freedley, *Philadelphia and Its Manufacturers*, 285–86.

28. Nathaniel A. Pratt, "The Origin of the Ashley River Phosphates, and their Comparative Solubility," (handwritten draft, 13 January 1872), Pratt Papers; Wines, *Fertilizer in America*, 117–18, 215n35; Pratt, "Present," 229. CMMC never produced fertilizer. Most companies incorporated with the legal ability to mine and manufacture but chose to do just one.

29. "The Phosphates of South Carolina as an Element of Wealth," *CDC*, 22 February 1869, 2.

30. J. Morgan, *Recollections*, 262–64; Iseley, Baldwin and Baldwin, *Plantations of the Low Country*, 39; Wines, *Fertilizer in America*, 215n35; *Drayton Hall Historic Structures Report*, 27–28.

31. "Memorandum of Agreement," 18 October 1867, and "Memorandum of Agreement," 18–20 October 1867, Pratt Papers; H.A.M. Smith, "Ashley River," 3–10; F. Holmes, *Phosphate Rocks*, 77.

32. "Memorandum of Agreement," 18 October 1867, and "Memorandum of Agreement," 18–20 October 1867, Pratt Papers. For more on mining restrictions, see CMMC legal document (10 March 1910) 14, MSP.

33. CMMC, "Articles of Agreement," 13 February 1868, Pratt Papers; CMMC, "Mortgage," 1 June 1894, MSP; CMMC document, 1910, SCHS; "Plan showing relative location of the Dorchester Road and the State Road . . . ," map (n.p., July 1911), SCHS; H.A.M. Smith, "Ashley River," 3–10.

34. CMMC, "Articles"; "Dorchester Road," map; H.A.M. Smith, "Ashley River," 3–10; "A Trip up Ashley River Banks," *CDC*, 2 July 1868, 2.

35. CMMC, "Articles." Mortgages secured the bonds so property owners could reclaim possession if CMMC failed to pay.

36. F. D., "The Charleston Phosphates," *CDC*, 25 December 1868, 2; CMMC, "Mortgage"; CMMC document, 1910, SCHS; "Minutes of Meetings Stockholders and Directors," 11, 22–23; Bob Lang, "Lakeside Living West Ashley Development Offers Homes on the Water," *CPC*, 13 June 1999, 1.

37. CMMC, "Articles"; Robert F. Hoke Papers, UNC–CH; Chapel Hill Iron Mountain Company Records, UNC–CH; Pratt, "Present," 227; Kerr, "Report on Cotton Production"; Kerr, "Geological Map of North Carolina"; Barefoot, *General Robert F. Hoke*, 1–10; J. A. Holmes, "A Sketch of Professor Washington Caruthers Kerr," Pamphlet, Rubenstein Library, Duke; Kerr, "North Carolina" cover page, 24–31; Winters, *Washington Caruthers Kerr*, 1–20.

38. "Charleston Mining and Manufacturing Company," *CDC*, 21 February 1868, 2; F. Holmes, *Phosphate Rocks*, page 4 of advertising section; Painter, "Recovery of Confederate Property," 82, 345, 391–92, 427, 430–31, 459; Wise, *Lifeline*, 222–24; Bulloch, *Secret Service*, 423–27; Nepveux, *Trenholm*, 96–98; Fraser, *Charleston!*, 281–83.

39. Williamson, *After Slavery*, 148–53; F. Holmes, *Phosphate Rocks*, 76–77.

40. Cheves, "Middleton of South Carolina," 250–51; Haskell, *Middleton Place Privy House*, 6–7.

41. J. F. Fisher (Philadelphia) to WM (Charleston), 8, 10, 11, 12, 14, 20, 22, 27 January 1868, WMP.

42. John I. Middleton Jr. (Baltimore and Waccamaw, S.C.) to WM (Charleston), 3, 4, 5, 25 February, 7 March 1868, and 16 April, WMP; "Constitution of the Ashley Mining & Phosphate Co. of South Carolina," [April 1868?], WMP.

43. "Phosphate and Marl Lands," *CDC*, 27 March 1868, 1; "Large Shipment of Phosphate Rocks," *CDC*, 22 April 1868, 2; "Charleston Phosphates," *SCIPL*, 76.

44. "Charleston Phosphates," *RC* (1869): 48; "Charleston Mining and Manufacturing Company," *CDC*, 10 May 1869, 1; "An Act to Ratify, Confirm and Amend . . . ," *Acts* 1868–70, 206–7.

45. "Large Shipment of Phosphate Rocks," *CDC*, 22 April 1868, 2; "The Phosphate Trade," *CDC*, 14 November 1868, 2; "The Charleston Phosphates," *CDC*, 25 December 1868, 2; F. Holmes, *Phosphate Rocks*, 75. Lamb's remained CMMC's base of operations for the next thirty years.

46. F. Holmes, *Phosphate Rocks*, 76; Chazal, *Century*, 49–51; Willis, "Marl Beds," 52, 60; "Substantial Enterprise," *CDC*, 30 November 1867, 2; "The Phosphates of the Ashley River," *CDC*, 16 December 1867, 2; Wines, *Fertilizer in America*, 118; "The Charleston Phosphate Beds," *CM*, 2 July 1868, 1.

47. "Substantial Enterprise," *CDC*, 30 November 1867, 2; Stephens, *Ancient Animals*, 41.

48. F. Holmes, *Phosphate Rocks*, 74; Simons & Locke Law Firm Records, SCHS; Census 1880 Population, Charleston County; "The Charleston Phosphate Beds," *CM*, 2 July 1868, 1.

49. "Remarkable Discovery," *CDC*, 5 October 1867, 2; "Another Remarkable Discovery," *CDC*, 8 October 1867, 2; "Substantial Enterprise," *CDC*, 30 November 1867, 2; CMMC ad, *CDC*, 17 December 1867, 3; "Interesting Geological Discoveries," *CDC*, 7 February 1868, 2; "Large Shipment of Phosphate Rocks," *CDC*, 22 April 1868, 2; "A Trip up Ashley River Banks," *CDC*, 2 July 1868, 2; "Phosphates," *CDC*, 30 July 1868, 2; "The Phosphate Trade of Charleston," *CDC*, 17 August 1868, 2; "The Phosphate Trade," *CDC*, 3 September 1868, 2; "The Phosphate Trade," *CDC*, 14 November 1868, 2; "The Charleston Phosphates," *CDC*, 25 December 1868, 2; "The Charleston Phosphate Beds," *CM*, 2 July 1868, 1; "The Phosphates of Ashley River," *CDC*, 16 December 1867, 2; "South Carolina Phosphates," *CDC*, 11 June 1868, 2; "The Phosphate Interest of Charleston," *CDC*, 29 October 1868, 2.

50. F. Holmes, *Phosphate Rocks*, 74 and ads; "A Trip up Ashley River Banks," *CDC*, 2 July 1868, 2; "The Charleston Phosphates," *CDC*, 25 December 1868, 2; "Charleston Mining and Manufacturing Company," *CDC*, 10 May 1869, 1; "The Charleston Phosphates," *RC* (1869): 48; "Charleston Mining and Manufacturing Company," *CDC*, 20 October 1869, 1.

51. Chazal, *Century*, 49; "There's Millions in It," Supplement to *CNC*, 1 March

1884, 1; "The Phosphates of the Ashley River," *CDC*, 16 December 1867, 2; "Charleston Phosphates," *RC* (1869): 48–49; Wines, *Fertilizer in America*, 117–19.

52. "Charleston Phosphates," *RC* (1869): 48; "Charleston Phosphates," *SCIPL*, 76–77. Although some farmers used ground, unprocessed phosphate as fertilizer in the early years, they soon grew to prefer the manufactured product.

53. Landes, *Unbound Prometheus*, 128–33, 528.

54. Edgar, *South Carolina*, 51, 182; Seymour, "Vive l'heritage Hugeno," 20–27; Poston, *Buildings of Charleston*, 89; "B. H. Rutledge," 26–27; Garlington, *Men of the Time*, 216.

55. "The Phosphates of the Ashley River," *CDC*, 16 December 1867, 2; Willis, "Marl Beds," 50–52; "Charleston Phosphates," *SCIPL*, 73; Chazal, *Century*, 46–47; "Remarkable Discovery," *CDC*, 5 October 1867, 2; "Another Remarkable Discovery," *CDC*, 8 October 1867, 2; "Substantial Enterprise," *CDC*, 30 November 1867, 2; CMMC ad, *CDC*, 17 December 1867, 3; "The Phosphate and Manufacturing Interests of South Carolina," *CDC*, 30 December 1868, 2.

56. "The Phosphates of the Ashley River," *CDC*, 16 December 1867, 2; "List of Prices," *CDC*, 22 November 1867, 4; Wines, *Fertilizer in America*, 68, 120, 175–76. A vertically integrated company owns parts of the supply chain in order to ensure delivery of raw materials for production.

57. Willis, "Marl Beds," 52; Wines, *Fertilizer in America*, 118; F. Holmes, *Phosphate Rocks*, 76; Chazal, *Century*, 49.

58. F. Holmes, *Phosphate Rocks*, 77–78.

59. Willis, "Marl Beds," 55–56; Chazal, *Century*, 49–51.

60. Chazal, *Century*, 49–51; "The Charleston Phosphate Beds," *CM*, 2 July 1868, 1; "A Trip up Ashley River Banks," *CDC*, 2 July 1868, 2.

61. Charles U. Shepard Jr., "South Carolina Phosphates and Their Principal Competitors in the Markets of the World," in *First Annual Report of the Commissioner of Agriculture, 1880*, 78–79, EWS-6.

62. Chazal, *Century*, 49–51; "Charleston Phosphates," *RC* (1869): 48; Hatch, "Contribution," 357.

63. "South Carolina Phosphates," *CDC*, 11 June 1868, 2; "The Wando Fertilizer," *CDC*, 18 September 1868, 2.

64. "Pacific Guano," *CDC*, 19 October 1868, 2; "The Phosphate Interest of Charleston," *CDC*, 29 October 1868, 2; "The Phosphates of South Carolina. The Pacific Guano Company," *CDC*, 31 August 1869; "The Charleston Phosphates," *RC* (1869): 48; "An Act to Incorporate the Wando Mining and Manufacturing Co.," *Acts 1868–70*, 67; Carlton, *Mill and Town*, 44–45; "Washington Light Infantry," *CDC*, 3 May 1869, 2.

65. "The Phosphate Interest of Charleston," *CDC*, 29 October 1868, 2; "The Charleston Phosphates," *RC* (1869): 48; "Works of the Wando Mining and Manufacturing Company," *RC* 2 (November 1870): 109–10; "Bird's Eye View" map; Wines, *Fertilizer in America*, 121.

66. Willis, "Marl Beds," 60; F. Holmes, *Phosphate Rocks*, 79; "There's Millions in It," *CNC*, 1 March 1884, 1. CMMC's totals rose from 6 tons (1867) to 4,383 (1868), 10,865 (1869), and 15,590 (1870). Wando resumed mining in later years.

67. F. Holmes, *Phosphate Rocks*, 51, 84–87, and ads; "Charleston Phosphates," *SCIPL*, 50, 76–79; "Charleston Phosphate Items," *RC* 1 (July 1870): 643–45; "Our Phosphate Beds," *CDC*, 25 August 1869, 4; Glover, *Narratives of Colleton County*, 163; Shepard Jr. "Report 1870," 8; Willis, "Marl Beds," 60–61, 65, 72–73; WM to: Dr. Rhett, 13 February 1869, John I. Middleton Sr., 21 February 1869, Charles Baker, 26 February 1869, and Eliza M. Fisher, 4 February 1870, WML; Chazal, *Century*, 54; Stono, *Almanac 1871*, 1, 16–19; "Bird's Eye View" map; H.A.M. Smith, map, 1; "Phosphate," *CDC*, 28 July 1868, 2; "Edisto Nature Trail."

68. F. Holmes, *Phosphate Rocks*, 51, 84–87, and ads; "The Phosphates of South Carolina. The Pacific Guano Company," *CDC*, 31 August 1869, 1; Peter B. Bradley to Frank E. Taylor, 30 March 1908, Frank E. Taylor Papers, USC; Wines, *Fertilizer in America*, 160; State of South Carolina v. the South Carolina Phosphate Company, Limited, alias the Oak Point Mines, Beaufort County, Court of Common Pleas, "Order of reference and report, with testimony" (Charleston: Walker, Evans & Cogswell, 1874), 94, SCHS.

69. F. Holmes, *Phosphate Rocks*, 84–87, and ads; "Charleston Phosphates," *SCIPL*, 50, 76–79; "Charleston Phosphate Items," 643–45; Comfort, "Correspondence," 407; WM to Charles Baker, 20 November 1869, WML, 271–78; "Our Phosphate Beds," *CDC*, 25 August 1869, 4; Census 1870 Population, Charleston County.

70. F. Holmes, *Phosphate Rocks*, 74–87, and ads; "Charleston Phosphates," *SCIPL*, 76–79; "Charleston Phosphate Items," 643–45.

71. Willis, "Marl Beds," 60, 78; Chazal, *Century*, 71.

72. Willis, "Marl Beds," 60; Nepveux, *Trenholm*, 90–91; H. Holmes, "The Trenholm Family," 158–59; Espenshade and Roberts, *Drayton Hall Tract*, 45; "Condensed Report of Drayton Hall and Mining Lease In Acct with F. H. Trenholm 1868," JDP; F. H. Trenholm to Dr. John Drayton, 11 November 1868, JDP.

73. Loftus C. Clifford and J. Fraser Mathewes, agents of Dr. John Drayton, to Moulton Emery and John T. Prentice, "Lease for three years," 16 January 1866, JDP; F. H. Trenholm to Dr. John Drayton, 11 November 1868, JDP; Clifford & Mathewes to Dr. John Drayton, 4 December 1868, JDP; Clifford & Mathewes to Dr. John Drayton, 15 December 1868, JDP; Mathewes to Dr. John Drayton, 15 March 1869, JDP.

74. Clifford and Mathewes to Emery and Prentice, "Lease for three years"; F. H. Trenholm to Dr. John Drayton, 11 November 1868, JDP; Clifford & Mathewes to Dr. John Drayton, 4 December 1868, JDP; Clifford & Mathewes to Dr. John Drayton, 15 December 1868, JDP; F. H. Trenholm to Dr. John Drayton, 11 January 1869, JDP; Mathewes to Dr. John Drayton, 15 March 1869, JDP; *Drayton Hall Historic Structures Report*, 27–29; Espenshade and Roberts, *Drayton Hall Tract*, 40, 45–47.

75. F. H. Trenholm to Dr. John Drayton, 11 January 1869, JDP; Mathewes to Dr.

John Drayton, 15 March 1869, JDP; "John Drayton and Charleston, S.C. Mining and Manufacturing Co. Lease," 26 January 1875, JDP; Willis, "Marl Beds," 60.

76. F. Holmes, *Phosphate Rocks*, 75; Clifford & Mathewes to Dr. John Drayton, 4 December 1868, JDP; Otto A. Moses, "Report of the State Inspector of Phosphates for the Year Ending December 3, 1872," in *Reports 1872–3*, 716–17; Willis, "Marl Beds," 60; "Bonds, Stocks, and Coupons," *CDC*, 1 September 1869, 4. CMMC appeared to be on solid financial footing, its stock trading at par on September 1, 1869.

77. Stephens, *Ancient Animals*, 47–48; "Phosphates," *CNC*, 14 August 1878, 2; Pratt, "Present," 227–29; Pratt, *Ashley River Phosphates*, 38–42; F. Holmes, *Phosphate Rocks*, 73; Shepard Jr., "Report 1870," 3–4; "Charleston Phosphates," *SCIPL*, 74–76; Willis, "Marl Beds," 69–70.

78. Fraser, *Charleston!*, 182; Edgar, *South Carolina*, 281; Coclanis, "Entrepreneurship and Economic History," 212; Coclanis, "The Rise and Fall of the South Carolina Low Country," 143–66.

79. G. Wright, *Old South, New South*, 113–14, 156–64.

80. Pratt, *Ashley River Phosphates*, 42.

Chapter 3. Land Miners and Hand Mining, 1867–1884

1. F. Holmes, *Phosphate Rocks*, 75; Tuten, *Lowcountry Time and Tide*, 6.

2. Kolchin, *American Slavery*, 200–201, 216–29.

3. Coclanis, *Shadow*, 133–37; Tuten, *Lowcountry Time and Tide*, 22–26, 39–46; Sass, "The Rice Coast," 13.

4. "Up the Ashley River," *RC* 2 (March 1871): 361.

5. Cothran, *Gardens of Historic Charleston*, 133–34; G. S. Rogers, "Phosphate Deposits," 185.

6. Shepard Jr., "South Carolina Phosphates. A Lecture," 12; Jennie Haskell, "A Visit to the Phosphate Fields and Hills," *Harper's Weekly*, 1885, 412, SCHS; Wyatt, *Phosphates of America*, 53; Waggaman, "Report," 6–7; Fuller, "History," 29; Chazal, *Century*, 18–19.

7. "The South Carolina Deposits of Bone Phosphate," *New York Times*, 4 June 1868, 5 (reprinted from the Macon, Georgia, *Telegraph*, 26 May 1868); F. Holmes, *Phosphate Rocks*, 7–14, 70–71; Shaler, *Phosphate Beds*, 11–12; Hammond, *South Carolina*, 48; Haskell, "Visit," 412; Chazal, *Century*, 9; Waggaman, "Report," 4–7, 9; G. S. Rogers, "Phosphate Deposits," 193, 213; "A Trip up Ashley River Banks," *CDC*, 2 July 1868, 2; "The Charleston Phosphate Beds," *CM*, 2 July 1868, 1; "The Charleston Phosphates," *RC* 1 (October 1869): 47; F. P. Porcher, "Popular View of South Carolina Phosphates," *CDC*, 18 August 1870, 2; "Early Notice of the Charleston Phosphates," *RC* 1 (July 1870): 640–41; "The Charleston Phosphates," *SCIPL*, 70–71, EWS-3.

8. Haskell, "Visit," 412; P. Morgan, "Work and Culture," 575; P. Morgan, "Task and Gang Systems," 195–97; Genovese, *Roll, Jordan, Roll*, 322–24; Heyward, *Mamba's Daughters*, 82–86. A careful researcher, Heyward wrote of paired miners (one digging and one wheelbarrowing) in the mines of the 1910s.

9. "The Charleston Phosphate Beds," *CM*, 2 July 1868, 1; "Bone Phosphate," *New York Times*, 4 June 1868, 5.

10. F. Holmes, *Phosphate Rocks*, 58–59; "There's Millions in It," Supplement to CNC, 1 March 1884, 1; G. S. Rogers, "Phosphate Deposits," 198–99, 209–11; Wyatt, *Phosphates of America*, 49–51, 54–56; Bowens, interview.

11. WM to Charles Baker (Baltimore), 11 October 1868, WML; Chazal, *Century*, 50. Working with steam shovels, later miners found much rock left by early hand miners.

12. Willis, "Marl Beds," 53; Tuomey, *Report* (1848), 53–55; F. Holmes, *Phosphate Rocks*, 58–59.

13. "The Charleston Phosphate Beds," *CM*, 2 July 1868, 1; "A Trip up Ashley River Banks," *CDC*, 2 July 1868, 2; "Our Phosphate Beds," *CDC*, 25 August 1869, 4; Haskell, "Visit," 412.

14. "The Phosphate Trade," *CDC*, 3 September 1868, 2; Waggaman, "Report," 7; G. S. Rogers, "Phosphate Deposits," 211, 214.

15. Tuomey, *Survey* (1844), 53–55; "The Phosphates of South Carolina as an Element of Wealth," *CDC*, 22 February 1869, 2; Moses, "Phosphate Deposits, 513–14; Hammond, *South Carolina*, 51; Day, "Fertilizers," 560; Waggaman, "Report," 7; Shepard Jr., "South Carolina. A Lecture," 12; Wines, *Fertilizer in America*, 118; Haskell, "Visit," 412; *South Carolina in 1884*, n.p.

16. Wyatt, *Phosphates of America*, 53; C. Wright, *Sixth Special Report*, 83–84; Waggaman, "Report," 7; Fuller, "History," 29; G. S. Rogers, "Phosphate Deposits," 210–11, 217; Millar, *Florida, South Carolina*, 153; Shepard Jr., "Report 1870," 9.

17. Haskell, "Visit," 411, 413; Wyatt, *Phosphates of America*, 53; Waggaman, "Report," 6–7; G. S. Rogers, "Phosphate Deposits," 210–11; Stephens, *Ancient Animals*, 38–44; F. Holmes, *Phosphate Rocks*, 9–12, 17, 35–41, 59; Tuomey, *Survey* (1844), 53–55; "The Charleston Phosphate Beds," *CM*, 2 July 1868, 1; "The Charleston Phosphates," *RC* 1 (October 1869): 47.

18. F. Holmes, *Phosphate Rocks*, 75; Haskell, "Visit," 411–13; WM to Baker: 20 June, 25 September, 11 October, and 4 December 1868, and 13 February, 19 September, and 29 October 1869, WML.

19. Shepard Jr., "Report 1870," 8; Willis, "Marl Beds," 60, 72, 78; Otto A. Moses, "Report of the State Inspector of Phosphates for the Year Ending December 3, 1872," *Reports 1872–3*, 720; Otto A. Moses, "Report of the Inspector of Phosphates to the General Assembly of South Carolina at Regular Session, 1873–4," (Columbia: Republican Printing Company, 1873), 8, EWS-5; Moses, "Phosphate Deposits," 513–14.

20. Haskell, "Visit," 411; "The Charleston Phosphate Beds," *CM*, 2 July 1868, 1; "A Trip up Ashley River Banks," *CDC*, 2 July 1868, 2; F. Holmes, *Phosphate Rocks*, 70; Moses, "Phosphate Deposits," 513; Chazal, *Century*, 9–10.

21. Kolchin, *American Slavery*, 31–32, 103–4, 106; Phillips, *American Negro Slavery*, 259; Coclanis, "How the Low Country," 61–63; P. Morgan, "Work and Culture," 564–66, 578; P. Morgan, "Task and Gang Systems," 191, 198–99; Engerman, "Economic

Response to Emancipation," 54; Stewart, "Rice, Water, and Power," 52; Saville, *Work of Reconstruction*, 7, 51–52.

22. P. Morgan, "Work and Culture," 199–202; Coclanis, "How the Low Country," 63–64; Hudson, *To Have and to Hold*, xix, 12.

23. Edgar, *South Carolina*, 313–14; Kolchin, *American Slavery*, 31, 103; Clifton, "The Rice Driver," 331, 336–39; P. Morgan, "Task and Gang Systems," 202–9; Phillips, *American Negro Slavery*, 247–48; P. Morgan, "Work and Culture," 581–82; Williamson, *After Slavery*, 136–37; Haskell, "Visit," 412; Bowens, interview.

24. Foner, *Nothing But Freedom*, 78–79; Hudson, *To Have and to Hold*, 16; P. Morgan, "Work and Culture," 566–69, 575–78; P. Morgan, "Task and Gang Systems," 204; Phillips, *American Negro Slavery*, 247–48; Coclanis, "How the Low Country," 76n38.

25. Schwalm, *Hard Fight*, 222.

26. Strickland, "Traditional Culture," 154; Hudson, *To Have and to Hold*, xix–xx, 15; P. Morgan, "Work and Culture," 566, 573–75, 579–80, 583–84, 590; P. Morgan, "Task and Gang Systems," 210–13, 218–20; Berlin, "Time, Space, and the Evolution," 60, 65–67; Coclanis, "How the Low Country," 61.

27. Dusinberre, *Them Dark Days*, vii–xi, 180, 253, 483–84; Kolchin, "Variable Institution," 111–21; Young, review, 400–401.

28. Strickland, "'Mud Work,'" 47; Strickland, "Traditional Culture," 142, 144–47, 163, 172n7–9; P. Morgan, "Work and Culture," 572–75, 579–81.

29. Hudson, *To Have and to Hold*, xx–xxii, 32, 182; Pruneau, review, 95–97; Kolchin, review, 1578–79; Strickland, "'Mud Work,'" 57; P. Morgan, *Slave Counterpoint*, 203; Hudson, review, 381–83; Dusinberre, *Them Dark Days*, viii, 410–14.

30. Kolchin, *American Slavery*, 216–29.

31. Schwalm, *Hard Fight*, 222; Saville, *Work of Reconstruction*, 66–67, 134–35; Rose, *Rehearsal for Reconstruction*, 82, 224–25; Woody, "Labor and Immigration Problem," 199.

32. P. Morgan, "Work and Culture," 584–85; Strickland, "Traditional Culture," 149; Schwalm, *Hard Fight*, 175–77, 183, 202–4.

33. Schwalm, *Hard Fight*, 156–57, 229.

34. Joyner, *Down by the Riverside*, 61; Edgar, *South Carolina*, 266–69.

35. P. Morgan, "Work and Culture," 585–86, 594–97; Saville, *Work of Reconstruction*, 115, 133; Schwalm, *Hard Fight*, 188, 204–7, 214–17, 226–232, 347n158; Williamson, *After Slavery*, 135–36; P. Morgan, "Task and Gang Systems," 219–20.

36. Williamson, *After Slavery*, 135–36; F. Holmes, *Phosphate Rocks*, 75.

37. Saville, *Work of Reconstruction*, 135–41; Shepard Jr., "Report 1870," 8.

38. *Census 1870 Population, Charleston County*; G. Wright, *Old South, New South*, 94–95; University of Virginia, *United States Historical Census Data Browser*. Fifty-nine percent of the state's population was black or mulatto, and 57 percent of all of its males were under the age of twenty-one.

39. *Census 1870 Population, Charleston County*; WM to RTS, 25 May 1868, WML; Shepard Jr., "Report 1870," 8; "Ashley River," *RC* 2 (March 1871): 361.

40. *Census 1870 Population, Beaufort, Charleston, and Colleton Counties*; Shepard Jr., "Report 1870," 8.

41. "Condensed Report of Drayton Hall and Mining Lease In Acct with F. H. Trenholm 1868," JDP; Saville, *Work of Reconstruction*, 179n113; "Element of Wealth," *CDC*, 22 February 1869, 2; WM to Baker, 23 June, 7, 22, 25 September, and 1 October 1868, WML.

42. Schwalm, *Hard Fight*, 204–14.

43. Jenkins, *Seizing the New Day*, 70–91, 171; Tindall, *South Carolina Negroes*, 215–19; Williamson, *After Slavery*, 236–39; Powers, *Black Charlestonians*, 118, 136–47; *Census 1870 Population, Charleston County*.

44. *Census 1870 Industry, Charleston, Beaufort, and Colleton Counties*; Willis, "Marl Beds," 60, 68, 72–73, 78; WM to Baker, 29 October 1869, WML; F. Holmes, *Phosphate Rocks*, 73–87; *Census 1870 Population, Beaufort, Charleston, and Colleton Counties*; Powers, *Black Charlestonians*, 249, 253. Data from Middleton Place corroborate this method: Ashley Mining employed thirty-five hands in October 1869; my estimated total for 1870 was thirty-four hands.

45. "The Charleston Phosphates," *CDC*, 25 December 1868, 2; "A Trip up Ashley River Banks," *CDC*, 2 July 1868, 2; "The Charleston Phosphate Beds," *CM*, 2 July 1868, 1; "Element of Wealth," *CDC*, 22 February 1869, 2; Willis, "Marl Beds," 60; "Charleston Phosphates," *SCIPL*, 76–79.

46. *Compendium Tenth Census*, lxii.

47. Schwalm, *Hard Fight*, 257–60, 263–68; Stanley, *From Bondage to Contract*, 35–37.

48. Schwalm, *Hard Fight*, 257–58.

49. Abbott, *Freedmen's Bureau*, 20–21.

50. *Census 1870 Population, Charleston and Colleton Counties*; "An Act to Grant to Certain Persons . . . the right to dig and mine in the beds of the navigable streams and waters of the state . . . ," *Acts 1868–70*, 381; *Compendium Ninth Census*, 610, 802, 869; Hammond, *South Carolina*, 578, 601.

51. *Census 1870 Population, Beaufort, Charleston, and Colleton Counties*; Saville, *Work of Reconstruction*, 135–41; "A Trip up Ashley River Banks," *CDC*, 2 July 1868, 2; "The Charleston Phosphate Beds," *CM*, 2 July 1868, 1; "The Phosphate Trade," *CDC*, 3 September 1868, 2.

52. *Compendium Ninth Census*, 838, 869–70; *Census 1870 Industry, Charleston County*, 732; *Compendium Tenth Census*, 929; Willis, "Marl Beds," 68; *Census 1870 Industry, Charleston County*; Shepard Jr., "Report 1870," 8.

53. Foner, *Reconstruction*, 77, 134–35; Scott, *Weapons*, xv–xvii; WM to John I. Middleton Sr. (Waccamaw), 6 December 1868, WML; WM to Edward Middleton, 1 July 1868, WML; Roark, *Masters Without Slaves*, 206.

54. WM to J. F. Fisher (Philadelphia), 12 November 1868, WML; WM to Baker, [number illegible] May and 8 May, 1 and 29 October 1869, WML; WM to Henry Middleton, 15 September 1870, WMP; John I. Middleton Jr. (Baltimore) to WM, 19

December 1870, WMP; WM to John I. Middleton Jr., 1 January 1871, WMP; Roark, *Masters Without Slaves*, 68–69, 106–11, 144–47, 152–59, 198–209; Lewis and Hardesty, "Middleton Place," 18.

55. WM to John I. Middleton Sr., 2 June 1868, WML; WM to Edward Middleton, 31 July 1868, WML; John I. Middleton Jr. to WM, 16 April 1868, WMP; Haskell, *Middleton Place Privy House*, 6–7.

56. WM to RTS, 25 May 1868, WML.

57. Schwalm, *Hard Fight*, 228–29; WM to John I. Middleton Jr., 1 January 1871, WMP.

58. WM to John I. Middleton Jr., 8 June 1868, WML; Abbott, *Freedmen's Bureau*, 20–21.

59. *Census 1870 Population, Charleston County*, 60–61, 258; Hammond, *South Carolina*, 678; Powers, *Black Charlestonians*, 100–103; Coclanis, *Shadow*, 112–15, 130, 276; *Compendium Tenth Census*, liv–lxv.

60. Stauffer, *The Formation of Counties*, 16–17; Barbara Doyle, Middleton Place, letter to author, 23 February 2001.

61. Foner, *Nothing But Freedom*, 37–38, 45, 90–91; Foner, *Reconstruction*, 138–40; WM to John I. Middleton Jr., 5 June 1868, WML; WM to Baker, 23 June 1868, WML.

62. WM to Baker, 26 July 1868, WML.

63. WM to Baker, 3 September 1868, 20 February 1869, and 18 March 1869, WML.

64. WM to RTS, 25 May 1868, WML; WM to Baker, 25 September 1868, WML; Roark, *Masters Without Slaves*, 165–69.

65. WM to Baker, 23 June 1868, 1 August 1868, 3 and 25 September 1868, 4 December 1868, 31 January 1869, [early May 1869?], [late July 1869?], 1 and 29 October 1869, WML.

66. WM to Baker, 23 June, 7 and 25 September, [1 October?] 1868, and 29 October 1869, WML.

67. Barbara Doyle, Middleton Place, letter to author, 14 September 2001; WM to RTS, 25 May 1868, WML; WM to John I. Middleton Jr., 31 May 1868, WML; WM to Baker, 25 May and 7, 25 September 1868, WML.

68. WM to Baker, 7, 22, 25 September, 5, 7 October, 21 November, and 4 December 1868, WML; WM to RTS, 22 October 1868 and 12 December 1869, WML; J. F. Fisher to WM, 20, 22, 27 January 1868, WMP; WM to John I. Middleton Sr., 6 December 1868, WML.

69. Ayers, *Promise of the New South*, 13–14; Schwalm, *Hard Fight*, 224–26; Foner, *Nothing But Freedom*, 87–90; WM to RTS, 25 May 1868, WML; WM to John I. Middleton Jr., 16 June 1868, WML.

70. WM to Eliza Fisher, 25 December 1868, WML; WM to Baker, 1 and 29 October 1869, WML; William S. Hastie Jr. to WM, 10 July 1873, WMP; WM to John I. Middleton Jr. (Baltimore), 6 October 1870, WMP; WM to J. Francis Fisher (Alverthorpe), 9 October 1870, WMP. He also suspected sabotage, reporting in October 1870 that a washer and engine had burned.

71. WM to Baker, 25 September 1868, WML; Baker to WM, 1 January 1869, BTL; RTS to WM, 22 August 1870, BTL; Shapiro, *New South Rebellion*, 238–41; Powers, *Black Charlestonians*, 127–34; Kolchin, *American Slavery*, 157–61; Scott, *Weapons*, xv–xvii.

72. John I. Middleton Jr. (Baltimore) to WM, 19 December 1870, WMP; WM to John I. Middleton Jr., 1 January 1871, WMP; Susan Middleton to Henry Middleton (Staunton, VA), 28 April 1872, WMP; Foner, *Nothing But Freedom*, 90–91; Powell, *New Masters*, xii–xiii, 3–7, 34, 52–53, 73, 97–99, 120–23, 155.

73. *Census 1880 Population, Charleston County*; G. Wright, *Old South, New South*, 94–95; University of Virginia, *United States Historical Census Data Browser*; Manning, *Black Apollo of Science*, 5, 14–17; Heyward, *Mamba's Daughters*, 82. Heyward included a woman miner in his novel but added that "the mines were for the men." The only other female land miner documented was Mary Just, who began working a "man's job" (gathering rocks) on James Island in 1887.

74. *Census 1880 Population, Charleston and Colleton Counties*.

75. *Census 1880 Population, Charleston and Colleton Counties*; Schwalm, *Hard Fight*, 204–7, 211–14, 235, 266–68, 272–73n10–12.

76. *South Carolina in 1884*, n.p.

77. C.R.M., "Digging Phosphate Rock," *New York Times*, 18 October 1881, 2; Espenshade and Roberts, *Drayton Hall Tract*, 47–54; Judd, "Roberts/McKeever," 2, 10.

78. Mancini, *One Dies, Get Another*, 208; "Annual Report of the Board of Directors and Superintendent of the South Carolina Penitentiary," *Reports 1880*, 5–6, 14; "Annual Report of the Board of Directors and Superintendent of the South Carolina Penitentiary," *Reports 1881–2*, 74–75, 82, 130; Fenhagen, "Descendants of Judge Robert Pringle," 306.

79. Espenshade and Roberts, *Drayton Hall Tract*, 47–54; Judd, "Roberts/McKeever," 2, 4, 8–10, 15–16; Judd, "Dennis, Washington and Nanny Notes," Housesites 1–5, 7, 13–14, 22–24, 26–31.

80. *Census 1880 Manufactures, Beaufort, Charleston, and Colleton Counties*; C.R.M., "Digging Phosphate Rock"; Moses, "Phosphate Deposits," 513–14; Hammond, *South Carolina*, 51; Haskell, "Visit," 412; *South Carolina in 1884*, n.p.; "Annual Report of the Commissioner of Agriculture of the State of South Carolina, 1885," *Reports 1885* II, 105; Tindall, *South Carolina Negroes*, 98n28; Williamson, *After Slavery*, 134–35.

81. *Census 1870 Population, Charleston County*; *Census 1880 Population, Charleston County*; Tindall, *South Carolina Negroes*, 215–19; Powers, *Black Charlestonians*, 118, 136–47; Jenkins, *Seizing the New Day*, 70–91, 171; Williamson, *After Slavery*, 236–39.

82. *Census 1880 Manufactures and Population, Beaufort, Charleston, and Colleton Counties*; "Piling on the Agony," CNC, [24?] September 1879, EWS-7; "Phosphates," *Reports 1879*, 103–12; State of South Carolina, County of Charleston, Summons for Relief, Catherine B. Smith v. The Wando Mining and Manufacturing Company, et al, 1880, SCHS; "Phosphate Mining," *Reports 1880*, 109–11, 172; E. L. Roche, "Annual Report," *Reports 1880*, 483–86; "Phosphate Mining in Beaufort," *Beaufort News*,

June 1880, reprint [*CNC*?], EWS-7; E. Willis, "Crude Phosphates," (10 March 1881), SCHS; Moses, "Phosphate Deposits," 519–20; Day, "Fertilizers," 786–87; *Compendium Tenth Census*, 1,014; Hammond, *South Carolina*, 50–51, 578, 601, 608, 665–68, 682.

83. *Census 1880 Population, Beaufort, Charleston, and Colleton Counties.*

84. Willis, "Crude Phosphates," (10 March 1881); E. L. Roche, "Phosphate Department," *Reports 1881–2*, 179–87; Moses, "Phosphate Deposits," 519–20; "The Transactions in Phosphate Rock, [1880–1882]," 1882, EWS-box; Day, "Fertilizers," 786–87; Hammond, *South Carolina*, 608; "South Carolina Phosphates," [*CNC*?] 1883, EWS-9; "There's Millions in It," *CNC*, 1 March 1884, 1; Chazal, *Century*, 71.

85. *Compendium Tenth Census*, 1,014. With women and children, cotton goods employed 2,018.

Chapter 4. River Mining and Reconstruction Politics, 1869–1874

1. Hammond, *South Carolina*, 56–58.

2. "To the Members of Senate and House of Representative of the Legislature of South Carolina," *CDC*, 16 December 1869, 2; "Legislative Proceedings," *CDC*, 16 December 1869, 4; "Legislative Proceedings," *CDC*, 20 December 1869, 4; "The Phosphate Ring," *CDC*, 21 December 1869, 1.

3. Edgar, *South Carolina*, 411. Before the 1886 general incorporation law, legislators granted corporate charters.

4. Bailey, Morgan, and Taylor, *Biographical Senate*, 1,570–72; *Journal of the Senate, 1869*, 110, 120; "Phosphates," *CDC*, 16 December 1869, 4; "Legislative Proceedings," *CDC*, 20 December 1869, 4; Coulter, *Williams*, 157–58.

5. *Journal of the House 1869–70*, 140, 152; Bailey, Morgan, and Taylor, *Biographical Senate*, 246–48, 401–2, 701–3, 817–18, 1,336–37, 1,482–83; "Legislative Proceedings," *CDC*, 20 December 1869, 4; Holt, *Black Over White*, 16, 114, 162; Williamson, *After Slavery*, 144, 204, 271, 280, 333, 337, 366–67, 379; Rosengarten, *Tombee*, 714; Zuczek, *State of Rebellion*, 160, 168–69; J. Reynolds, *Reconstruction*, 226–29; State v. the South Carolina Phosphate Company, Limited, alias the Oak Point Mines, Order of Reference and Report, with Testimony, Beaufort County Court of Common Pleas (Charleston: Walker, Evans & Cogswell, 1874), 54–55, SCHS. The resulting company hired Wells as mining superintendent.

6. "To the Members . . . ," *CDC*, 16 December 1869, 2.

7. "To the Members . . . ," *CDC*, 16 December 1869, 2; "Legislative Proceedings," *CDC*, 16 December 1869, 4.

8. "The State and Its Right Over the Phosphates Under the Sea," *CDC*, 17 December 1869, 2; "The Phosphate Question," *CDC*, 23 December 1869, 2; "The Phosphate Bill," *CDC*, 27 December 1869, 2.

9. "The Phosphate Question," *CDC*, 23 December 1869, 2; *Journal of the Senate 1869*, 187; "The Phosphate Ring," *CDC*, 21 December 1869, 1; Fraser, *Charleston!*, 288–89; Williamson, *After Slavery*, 227, 285, 338; Powers, *Black Charlestonians*, 120–21, 132,

161–62; Zuczek, *State of Rebellion*, 76. Charleston had an oversupply of labor, but mining regions and the rest of the state did not.

10. Bailey, Morgan, and Taylor, *Biographical Senate*, 327–29; "Legislative Proceedings," *CDC*, 16 December 1869, 4; "To the Members . . . ," *CDC*, 16 December 1869, 2; "Legislative Proceedings," *CDC*, 20 December 1869, 4; Williamson, *After Slavery*, 364.

11. Williamson, *After Slavery*, 370; Bailey, Morgan, and Taylor, *Biographical Senate*, 1,191–93; "Legislative Proceedings," *CDC*, 20 December 1869, 4; *Journal of the Senate 1869*, 152–53, 177; "Our Columbia Correspondence," *CDC*, 23 December 1869, 4; "Legislative Proceedings," *CDC*, 24 December 1869, 1. Willis lobbied for the bill but was not initially part of the list.

12. "Our Columbia Correspondence," *CDC*, 23 December 1869, 4; "Legislative Proceedings," *CDC*, 24 December 1869, 1; "An Act to Incorporate the South Carolina Phosphate Company," *Acts 1868–70*, 187–89; Stono, *Almanac, 1871*, 1; J. F. Fisher (Philadelphia) to WM, 20 January 1868, WMP; H, "The Phosphate Companies. Some Reasons to Think the Taxes Should be Remitted," Letter to the Editor, *CNC*, 17 December 1879, EWS-7; Chazal, *Century*, 71; Coulter, *Williams*, 175; Foner, *Reconstruction*, 466–67.

13. "Legislative Proceedings," *CDC*, 24 December 1869, 1; "The Phosphate Bill," *CDC*, 27 December 1869, 2; Edgar, *South Carolina*, 283; Fraser, *Charleston!*, 208.

14. Downey, "Riparian Rights," 84–87, 90–108.

15. Downey, "Riparian Rights," 104.

16. Hahn, *Roots of Southern Populism*, 58–63, 239–68; Edgar, *South Carolina*, 411.

17. Geisst, *Monopolies*, 6, 11–18, 31; Ritter, *Goldbugs and Greenbacks*, 2–8, 30; Wiebe, *Search for Order*, 7–8, 45–46, 52–53; Foner, *Reconstruction*, 22–23; Edgar, *South Carolina*, 282–84. Natural monopolies were capital-intensive, often extractive, and strategically important industries, in which, some claimed, competition was unrealistic and regulation socially harmful.

18. Geisst, *Monopolies*, 18–21, 23–26, 31; Fraser, *Charleston!*, 311; Williamson, *After Slavery*, 383–87.

19. Free Trade, "The Phosphate Outrage. A Letter to Corbin," *CDC*, 11 January 1870, 2; "South Carolina Legislature," *CDC*, 18 January 1870, 4; Foner, *Reconstruction*, 412–21, 572–73; Edgar, *South Carolina*, 402–3; "The Phosphate Deposits in Our Navigable Streams," *CDC*, 18 January 1870, 2.

20. "Our Columbia Correspondence," *CDC*, 5 January 1870, 1; Zuczek, *State of Rebellion*, 71–78; "Legislative Proceedings," *CDC*, 11 January 1870, 1; "South Carolina Legislature," *CDC*, 12 January 1870, 4; Williamson, *After Slavery*, 358; *Journal of the Senate 1869*, 3, 210; Bailey, Morgan, and Taylor, *Biographical Senate*, 75, 246–47, 328, 702–3, 921–22; Reynolds and Faunt, *Biographical Senate*, 62.

21. *Journal of the Senate 1869*, 3; "South Carolina Legislature," *CDC*, 12 January 1870, 4; "South Carolina Legislature," *CDC*, 18 January 1870, 4; Williamson, *After Slavery*, 330; "From Columbia," *CDC*, 20 January 1870, 2; H. Richardson, *Death of*

Reconstruction, 6–7, 89–92, 119–20; Fraser, *Charleston!*, 289. Tax reforms in 1868 targeted landowners, and land reform included state funds for black settlers.

22. H. Richardson, *Death of Reconstruction*, 6–7, 89–92, 119–20.

23. "From Columbia," *CDC*, 18 January 1870, 4; "From Columbia," *CDC*, 2 February 1870, 2; Williamson, *After Slavery*, 358–62; Kolchin, "Scalawags, Carpetbaggers, and Reconstruction," 67–76; Holt, *Black Over White*, 124, 143–44, 148–50; Zuczek, *State of Rebellion*, 75–83; Foner, *Reconstruction*, 389.

24. Bailey, Morgan, and Taylor, *Biographical Senate*, 762–63; "South Carolina Legislature," *CDC*, 18 January 1870, 4; Williamson, *After Slavery*, 146; *Journal of the House 1869–70*, 209, 225, 447–48; Holt, *Black Over White*, 229; *Journal of the Senate 1869*, 215, 232–33, 301–2, 305, 317, 340, 344–50; "South Carolina Legislature," *CDC*, 28 January 1870, 1; "The Phosphates," *CDC*, 2 February 1870, 2; Coulter, *Williams*, 161; "The Phosphates of South Carolina as an Element of Wealth," *CDC*, 22 February 1869, 2; "The Phosphates," *CDC*, 5 February 1870, 4.

25. "From Columbia," *CDC*, 3 March 1870, 4; *Journal of the Senate 1869*, 350, 517–20, 526; *Journal of the House 1869–70*, 516; Edgar, *South Carolina*, 387, 394; "From Columbia," *CDC*, 4 March 1870, 2.

26. "From Columbia," *CDC*, 4 March 1870, 2; "From Columbia," *CDC*, 18 January 1870, 4; "Legislative Proceedings," *CDC*, 20 December 1869, 4; "Our Columbia Correspondence," *CDC*, 23 December 1869, 4; "Another Committee Reports," (Beaufort) *Tribune and Commercial*, 10 October 1878, EWS-7; J. Reynolds, *Reconstruction*, 126, 487; Wallace, *The History of South Carolina*, 265; *Report Insurrectionary States*, vol. 1, 87–88, 122, and vol. 4, 68–89, 725, 729, 736, 748; Coulter, *Williams*, frontispiece, 109, 157–62, 228, 258; Conner, "Report," 42, EWS-7; Williamson, *After Slavery*, 376, 387, 394; "Phosphates," *Reports 1877–78*, 1,676–78. Williamson argues that railroad and printing schemes initiated the peak of Reconstruction-era corruption (1870–74), but the phosphate bill predated both. Redeemers later blamed Hurley and other Republicans for the phosphate legislative fiasco while ignoring the roles that Democrats and allies—such as Willis and Williams—played as lobbyists.

27. *Journal of the Senate 1869*, 353, 365–66, 381–82, 389, 395, 460, 500; "An Act to Grant to Certain Persons Therein Named . . . ," *Acts 1868–70*, 381; Coulter, *Williams*, 23, 61–64, 70, 83, 112, 116–18, 126–34, 135–50, 156–62, 169, 172, 260; Willis, "Marl Beds," 72–73; Frank E. Taylor will (1912), Frank E. Taylor Papers, Duke; Frank E. Taylor Papers, USC; Coker, *Maritime*, 278–83; Wise, *Lifeline*, 221; Lander, "Charleston," 346, 349.

28. William L. Bradley v. The Marine and River Phosphate Mining & Manufacturing Co. of South Carolina, Louis D. Mowry, and others, Bill of Complaint, United States District of South Carolina, Fourth Circuit, 1877, SCHS; Wines, *Fertilizer in America*, 109, 120, 160, 210n78; Willis, "Marl Beds," 60–61; Taylor Papers, USC; Jacob, "History and Status," 42–43; Coulter, *Williams*, 153, 161; William L. Bradley v. The South Carolina Phosphate and Phosphatic River Mining Company in South

Carolina, et al, Statement of Facts, U.S. District Court (Fourth Cir., 1874), 18–19, SCHS; Carolina Fertilizer ad, *CDC*, 1 December 1869, 3.

29. The Marine and River Phosphate Mining & Manufacturing Co. of South Carolina, *Statute Law, Organization and By-Laws, Opinion of Hon. A. J. Willard* (Charleston: Walker, Evans & Cogswell, 1871), 10–11, EWS-2; "Exclusive Right Phosphate Bill," *CDC*, 26 March 1870, 1; Marine, *By-Laws 1870*, 2, 4–7; Bradley v. SCPP, Statement of Facts (1874), 13–15; Bradley v. Marine, Bill of Complaint (1877), 1, 4–7; Coulter, *Williams*, 162; Shepard Jr., "Report 1870," 9; "General Table showing Total Shipments of Rock by every Company or Individual Working under Grant, Charter or License from the State since 1870," courtesy of Mary Miller, [*Reports 1887?*], n.p. Marine's presidents included Williams (1870) and Corbin (1871). Investors included Hurley, Willis, Chisolm, and future governors Daniel Chamberlain and Reuben Tomlinson.

30. Charles U. Shepard Sr., "On the Phosphatic Sand of South Carolina," *RC* 2 (May 1871): 438–41; Willis, "Marl Beds," 53; "The Wealth in Phosphates: The Developing of the Mining Business," [*CNC*? 1880?] EWS-10; F. Holmes, *Phosphate Rocks*, ads.

31. "That Phosphate Business," *CDC*, 25 June 1870, 2; Bradley v. SCPP, Statement of Facts (1874), 49, 107–23; Julie Powers Bradley, e-mails to author, 2 February 2000, 26 March 2002; "The Phosphate Cases," *CDC*, 20 July 1870, 2; "From Columbia," *CDC*, 30 July 1870, 1; Marine, *By-Laws* 1871, 18; Coulter, *Williams*, 161; "From Columbia," *CDC*, 14 January 1871, 4; "New Phosphate Company," *CDC*, 16 January 1871, 4; *Journal of the House 1870–71*, 142–43, 148, 196–97, 216, 344, 435, 478, 536–37, 540, 568, 576–77, 591–92; Faunt and Rector, *Biographical House*, 407–15; Holt, *Black Over White*, 229–41; Hine, "Black Politicians," 128–41; *Journal of the Senate 1870*, 390–91, 400, 415, 426–27, 430; Bailey, Morgan, and Taylor, *Biographical Senate*, 701, 1,482–85, 1,718–20; Seip, *South Returns to Congress*, 80–82; "A Carpetbagger in Disgrace," *CDC*, 25 February 1870, 1; "An Act to Charter the South Carolina Phosphate and Phosphatic River Mining Company," *Acts 1871–1872*, 688–89; Reynolds and Faunt, *Biographical Senate*, 63.

32. "South Carolina Phosphate and Phosphatic," *Acts 1871–72*, 688–89; Conner, "Report," 8, 54; Hine, "Dr. Benjamin A. Boseman Jr.," 346–48; Foner, *Freedom's Lawmakers*, 143; Powers, *Black Charlestonians*, 169–70; "Legislative Proceedings," *CDC*, 20 December 1869, 4; Williamson, *After Slavery*, 145, 271, 280, 333; Coulter, *Williams*, 161; *Journal of the Senate 1869*, 350, 526; *Journal of the House 1869–70*, 140, 152, 447, 516; *Report Insurrectionary States* v.1, 539; Holt, *Black Over White*, 59–60, 102–3, 124, 131–32, 143–44, 146, 148–50, 155, 159, 164–65, 197–98, 213. Legislators kept SPCC's royalty, license fee, and bond identical to Marine's.

33. SCPP, *By-Laws*; Bradley v. SCPP, Statement of Facts (1874), 17–18, 20–21, 24, 37–38; J. Reynolds, *Reconstruction*, 226; "Major Willis Dead," *CNC*, 1 March 1910, SCHS; Conner, "Report," 8–9, 13–14, 37, 54–55; "Legislative Proceedings," *CDC*,

20 December 1869, 4; "Commission Merchants" and "Factors," *Charleston Directory 1872–73*, 252, 254, 260.

34. Robert Smalls and N. B. Myers, "Report of the Special Joint Committee to Investigate Mining and Removing of Phosphates, &c.," *Reports 1871–2*, 996–99, 1,064; *Journal of the Senate 1871*, 499; *Journal of the Senate 1871–2*, 195, 496–99; Marine *By-Laws* 1870, 2–7; Marine *By-Laws* 1871, 2; "Act to Provide for the Appointment of an Inspector of Phosphates, and to declare his Duties," *Acts 1871–72*, 104–6; Otto A. Moses, "Report of the State Inspector of Phosphates for the Year Ending December 3, 1872," *Reports 1872–3*, 715–16; Otto A. Moses, "Report of the Inspector of Phosphates to the General Assembly of South Carolina at Regular Session, 1873–4," *Reports 1873–4*, 507–8; J. Reynolds, *Reconstruction*, 222–26; "Phosphate Companies," *Charleston Directory 1872–73*, 363; "'The Phosphate Case" [CNC?] n.d., EWS-5; "'The Phosphate Case," CNC, n.d., EWS-5; Bradley v. SCPP, Statement of Facts (1874), 18–21; "Act to Charter the Boatmen's Phosphate River Mining Company . . . ," *Acts 1873–74*, 530–31; William L. Bradley v. The South Carolina Phosphate and Phosphatic River Mining Company in the State of South Carolina, and the Marine and River Phosphate Mining and Manufacturing Company of South Carolina, Bill of Complaint for Injunction of Relief, U.S. District Court, In Equity (1873), 4–10, EWS-5; "A Harbor Master and the Phosphates," *CDC*, 24 February 1871, 2.

35. Raymond, "Historical Sketch of Mining Law," 988–1,004; Bradley v. SCPP, Statement of Facts (1874), 20, 22. Existing U.S. mining laws concerned mining on or below land, leaned toward laissez faire more than regulation, and ignored mining efficiency.

36. Shepard Jr., "Report 1870," 8; Bradley v. SCPP, Statement of Facts (1874), 20, 22; Shick and Doyle, "Boom," 20–21.

37. Bradley v. SCPP, Statement of Facts (1874), 20–21, 23–44, 47–56, 61–62, 67–69, 77–79, 88, 90, 95–97, 99, 105–23, 125–33; "General Table since 1870," (1888); "'That Phosphate Business," *CDC*, 25 June 1870, 2; Julie Powers Bradley e-mails to author, 2 February 2000, 26 March 2002. The barge-like (thirty-inch draft) dredge carried a crane, beam, and dipper that worked a twenty-four-foot circumference. Operators "threw" the dipper out and down to the river bottom and then dumped a cubic yard of material into a washer (four-foot draft). Lighters hauled away cleaned rock.

38. Moses, "Inspector 1872," 716; Bradley v. SCPP, Statement of Facts (1874), 15; Willis, "Marl Beds," 60; "General Table since 1870" (1888); Conner, "Report," 4–9, 26–27, 34, 48, 56; Robert Adger et al v. Farmers' Phosphate Company, S.C. Supreme Court, First Circuit, Charleston County, Statement of Case and Exceptions on Appeal (Charleston: News and Courier Book Presses, 1879), 12–13, MSP; Chazal, *Century*, 53; Coulter, *Williams*, 161–62.

39. Adger v. Farmers,' Statement of Case (1879), 9–10, 12–14; State of South Carolina v. Pacific Guano Company, S.C. Supreme Court (April Term, 1884), 2–4, MSP; Sanford, *Progression of Coosaw Plantation*, 5; Conner, "Report," 34–36; Coosaw Mining Company v. The Farmers' Phosphate Company, Testimony Introduced

by Plaintiffs at the Trial of the Case in Charleston, Charleston County Court of Common Pleas (6 March 1879), 69–76, 84, 99–100, 105, SCHS; The Coosaw Co. Minutes, Annual Meeting 27 June 1907, SCHS; Poston, *Buildings of Charleston*, 56, 409, 473, 549; Rosen, *Jewish Confederates*, 46, 143, 382n23; Hagy, *This Happy Land*, 357–58; "Commission Merchants" and "Copartnership Directory," *Charleston Directory 1872–73*, 253, 287; Bradley v. SCPP, Statement of Facts (1874), 47–53; Moses, "Inspector 1872," 715–16; Moses, "Inspector 1873," 507–8.

40. Conner, "Report," 34–35; Robert Adger (Charleston) to James Conner [Columbia?], 31 August 1877, 1, SCHS; A. Moore, *Biographical Directory*, 2–3; "Death of Mr. Robert Adger," [*CNC*? 9 May 1891?], Adger Family Papers, SCHS; Azrael, *The Story of Alex. Brown & Sons*, 58, 128–29, 141, 255–56; Killick, "The Cotton Operations of Alexander Brown," 170; Perkins, *Financing Anglo-American Trade*, 96–97, 101; Adger, *My Life and Times*, 36–37; Adger v. Farmers,' Statement of Case (1879), 13–14; Robert Adger, Joseph E. Adger and others, partners in a Joint Stock Company, and Mining Partnership, under the name and style of The Coosaw Mining Company, Plaintiffs, against The Farmers' Phosphate Company, a body politic and corporate under the laws of the State of South Carolina, Charleston County Court of Common Pleas, Decree, 15 March 1879, 3–4, SCHS.

41. F. Holmes, *Phosphate Rocks*, 81, ads; Willis, "Marl Beds," 60–61, 72–73; Bradley v. SCPP, Statement of Facts (1874), 13–15; Conner, "Report," 6–8; Thomas Taylor, "Report of the Inspector of Phosphates," *Reports 1879*, 107–8; Tischendorf, "A Note on British Enterprise," 196–97; "Retirement of Col. George W. Scott," 120; Land, *Charleston*, 151; Armes, *Story of Coal*, 330–32; Wilkins, "Free-Standing Company," 259–61, 264–70, 273–74. By 1914, free-standing companies comprised almost a third of all British overseas investments; others followed Wyllie Campbell into the South.

42. Bradley v. SCPP, Statement of Facts (1874), 107; State v. the South Carolina Phosphate Company, alias the Oak Point Mines, Order of Judge John J. Maher, Beaufort County Court of Common Pleas (December 1875), 1–6, SCHS; Willis, "Marl Beds," 60; "General Table since 1870," (1888); Conner, "Report," 9–10, 28–29, 52–54; Wyatt, *Phosphates of America*, 55; John B. Martin (Martin's Bank Ltd., London), to Mitchell & Smith (law firm, Charleston), 17 March 1894, MSP; Taylor, "Inspector 1879," 105–8; "The Special Joint Committee to Investigate Mining and Removing of Phosphates, &c," *Reports 1871–72*, 996–98; "The State v. The South Carolina Phosphate Company Limited, alias 'The Oak Point Mines,'" *Reports 1875–6*, 370; "Report of the Committee of the Senate Appointed to Investigate the Conduct of David T. Corbin in his Efforts to be Elected to the United States Senate," *Reports 1877–8*, 951–56, 964–65, 984–85, 990, 993–94; State v. Oak Point Mines, "Order of Reference," 1–125, SCHS; State v. the South Carolina Phosphate Company Limited, alias the Oak Point Mines, Report of Asher D. Cohen, Referee, Beaufort County Circuit Court of Common Pleas (Charleston: Walker, Evans & Cogswell, 1875), 1–16, SCHS; J. Reynolds, *Reconstruction*, 434–35, 461; "Act to Charter the Boatmen's . . . ," *Acts 1873–74*, 530–31. Boatmen's incorporators included Prince F. Stevens, July

Rivers, and John M. Freemen Jr., names that suggest that they were black, possibly former slaves.

43. Willis, "Marl Beds," 53; McGuire, "Getting Their Hands on the Land," 169–72.

44. Kiser, *Sea Island to City*, 89–92; Rupert S. Holland, ed., *Letters and Diary of Laura M. Towne*, typescript, 379–81, 427–33, 438–39, 453–54, 484–85, Penn School Papers; Griswold, *Sea Island*, 212–13.

45. "Mutiny at Bull River," *Beaufort Republican*, 5 December 1872, 2; "Drowning of Twelve Men," *Port Royal Commercial*, 20 November 1873, 3; Bradley v. SCPP, Statement of Facts (1874), 22, 73.

46. Coosaw v. Farmers,' Testimony (1879), 74, 77–78, 84, 87–90, 94, 96; Kiser, *Sea Island to City*, 89–92; Cecelski, *Waterman's Song*, 213–20.

47. *Census 1880 Manufactures, Beaufort and Charleston Counties*; C. Wright, *Sixth Special Report*, 130–31, 137; *South Carolina in 1884*, n.p.; "Annual Report of the Commissioner of Agriculture of the State of South Carolina, 1885," *Reports 1885* II, 105; Carlton, *Mill and Town*, 85–86n7. Both firms listed ten-hour workdays, but Coosaw's crews likely worked longer when necessary. Average wages for the state's male farm laborers was forty-five cents per day (1885), and textile wages (1883) were slightly more.

48. Conner, "Report," 35–36.

49. "General Table since 1870," (1888); F. Holmes, *Phosphate Rocks*, 81–83; Charles U. Shepard Jr., "South Carolina Phosphates and Their Principal Competitors in the Markets of the World," 83–84, EWS-6; Smalls and Myers, "Report," 999; Conner, "Report," 34; Coosaw Mining Company, Complainants, v. The Carolina Mining Company, et al, Defendants, Bill for Injunction and Relief, United States District Court-In Equity (Charleston: Walker, Evans & Cogswell, 1891), 7–8, SCHS; Millar, *Florida, South Carolina*, 159; Chazal, *Century*, 55; "Phosphates," *Beaufort Republican*, 24 April 1873, 2.

50. Wickwar, *300 Years*, 113–22; "Legislative Proceedings," *CDC*, 20 December 1869, 1; Chazal, *Century*, 51, 59, 70–71. Europeans preferred river over land rock.

Chapter 5. Convergence and the Fertilizer Industry, 1868–1884

1. Edward Willis, "Phosphates and Phosphatic Fertilizers," (31 January 1882), EWS-6.

2. Wines, *Fertilizer in America*, 41–42, 109–11, 126–32, 157–59; Thomas J. Moore, "Does Guano Pay?" *RC* 2 (May 1871): 447–49; Shepard Jr., "Report 1870," 3–7; "An Act to Regulate the Manufacture and Sale of Commercial Fertilizers . . . ," and "An Act to Provide for the Appointment of an Inspector of Phosphates . . . ," *Acts 1871–72*, 33, 104–6; Charles U. Shepard Jr., "Report of Charles U. Shepard Jr., Inspector of Guano and Fertilizers, to the General Assembly of South Carolina, November 1871," *Reports 1871–72*, 247–49; Otto A. Moses, "Report of the State Inspector of Phosphates for the Year Ending December 3, 1872," *Reports 1872–73*, 717–20.

3. F. Holmes, *Phosphate Rocks*, 73–74, 78–81, 84–87, ads; Shepard Jr., "Report 1870," 5–6; "An Act to Incorporate the Charleston Fertilizer Company," *Acts 1873–74*,

681; Thomas Taylor, "Statement of Fertilizers Inspected by Thos. Taylor for one year, from September 1st, 1878, to August 30th, 1879," *Reports 1879*, 121–22; "Table of Analyses, Commercial Values and Prices Per Ton, etc. of Commercial Fertilizers, Inspected, Analyzed and admitted to Sale in Georgia, for the Season of 1877–78," 4–10, EWS-7; Edward Willis, "Phosphates, Crude and Manufactured," (31 January 1883) EWS-9.

4. Willis, "Phosphates," (1883); E. L. Roche, "Phosphate Department," *Second Annual Report of the Commissioner of Agriculture 1881*, 15, EWS-6; "The Phosphate Business," [*CNC*?] 1882, EWS-9; Moses, "Phosphate Deposits, 520; "Charleston in 1883," *CNC*, 1883, EWS-9; "South Carolina Phosphates: Mining the Rock-Manufacturing Fertilizers," [*CNC*?] 1883, EWS-9; Day, "Manufactured Fertilizers," 825–26; South Carolina Department of Agriculture, "Analyses and Commercial Values of Commercial Fertilizers and Chemicals," 1884–85, (1885), 1–16, Hinson Pamphlets, Charleston Library Society.

5. F. Holmes, *Phosphate Rocks*, 73–74, 78–81, 84–87, ads; Willis, "Phosphates," (1883).

6. Wines, *Fertilizer in America*, 117–24; Gibbes, "Report," 1–6, 11–12; W. Ravenel, "Reply," 9.

7. Arthur M. Huger, "To the Stockholders of the Stono Phosphate Company," 2–8, 16–17, SCHS; Stono Phosphate Company, "Reply of President and Board," 12–13; W. Ravenel, "Reply," 9–11, 20–23; Stono Phosphate Company, *Revised By-Laws*, 3–6, SCHS; Stono, *Almanac 1880*, 1–6.

8. Jowitt, *Jowitt's Illustrated*, 42–52, 252–60; Wando Fertilizer ad, *CDC*, 16 January 1869, 3; Lander, "Charleston," 333–34.

9. Union Bank of South Carolina v. The Wando Mining and Manufacturing Company, Summons and Complaint for Relief, Charleston County Court of Common Pleas (Charleston: Walker, Evans, & Cogswell, 1878), 6–11, SCHS; G. C. Rogers, *Lawyers*, 70, 72, 77–78, 80–82, 312n87; "Necrology," 195–96; Reynolds and Faunt, *Biographical Senate*, 307; Fraser, *Charleston!*, 292; O'Brien, *Legal Fraternity*, 79–80, 91–92.

10. "An Act to Incorporate the Atlantic Phosphate Company," *Acts 1868–70*, 67; F. Holmes, *Phosphate Rocks*, ads; Atlantic, *Almanac 1872*, n.p.; Willis, "Marl Beds," 72–73; "Phosphate Companies, Agencies, &c.," *Charleston Directory 1872–73*, 363; "An Act to Incorporate the Pelzer Manufacturing Company," *Acts 1880*, 337; Land, *Charleston*, 145; Carlton, *Mill and Town*, 44–46; Lesser, *Relic of the Lost Cause*, 22–27; Mazyck, *Guide*, 169, 171.

11. "An Act to Amend the Charter of the Wando Mining and Manufacturing Co.," *Acts 1875–6*, 34–35; Union v. Wando (1878), 6–11; Catherine B. Smith v. The Wando Mining and Manufacturing Company, Summons for Relief, County of Charleston Court of Common Pleas (1880), 4–9, SCHS; Wando, *Almanac 1872*, n.p.

12. "The Phosphates of South Carolina. The Pacific Guano Company," *CDC*, 31 August 1869, 1; Willis, "Marl Beds," 72–73; Pacific Guano Company, *Pacific Guano*

Company, frontispiece, 7–11, 41; Wines, *Fertilizer in America*, 79–82; Shaler, *Phosphate Beds*, 12–13; "Pacific Guano," *CDC*, 19 October 1868, 2.

13. John B. Sardy ad, *CDC*, 2 March 1868, 3; *Wilson's Business Directory of New York City*; Wines, *Fertilizer in America*, 121–22; "Assignee's Sale," *CDC*, 24 November 1868, 3; Payne, "Plan of Wappoo"; Shepard Jr., "Report 1870," 5; "Bird's Eye View" map; Willis, "Marl Beds," 79.

14. John B. Sardy ad, *CDC*, 3 March 1871, 3; *Ashepoo Prospectus*, 5, EWS-5; F. Holmes, *Phosphate Rocks*, 51, 87, ads.

15. "Enormous Bones," *CDC*, 17 March 1870, 1; Carolina Fertilizer ad, *CDC*, 25 February 1869, 3; Wines, *Fertilizer in America*, 120, 134–35; "Commercial Review of the Past Year," *CDC*, 1 September 1869, 4; F. Holmes, *Phosphate Rocks*, 84, ads; "The Charleston Phosphates," *RC* 1 (October 1869): 48–49; Willis, "Marl Beds," 72–79; Shepard Jr., "Report 1870," 6. Shepard was an exception.

16. "Home Productions and Home Industry," *CDC*, 30 May 1868, 2; "The Phosphate and Manufacturing Interests of South Carolina," *CDC*, 30 December 1868, 2.

17. Wines, *Fertilizer in America*, 120–21; "Home Productions and Home Industry," *CDC*, 30 May 1868, 2; "The Charleston Phosphates," *CDC*, 25 December 1868, 2; "The Phosphate and Manufacturing Interests of South Carolina," *CDC*, 30 December 1868, 2; Coker, *Maritime*, 187–90.

18. Sulphuric Acid and Super-Phosphate Co. ads, *CDC*, 28 December 1868, 4, and 2 December 1869, 3; F. Holmes, *Phosphate Rocks*, ads.

19. Willis, "Marl Beds," 79; King, *Great South*, 450.

20. Willis, "Marl Beds," 72–79; Day, "Fertilizers," 818–19.

21. Willis, "Marl Beds," 72–79; "The Phosphate Interest of Charleston," *CDC*, 29 October 1868, 2; "The Phosphate Interests of Charleston," *CDC*, 9 February 1869, 2; "Bird's Eye View" map; Fraser, *Charleston!*, 208, 233, 304; "Map of Charleston" in *Year Book—1883*.

22. "The Storing of Guano in the City," *CDC*, 31 March 1869, 1.

23. Williamson, *After Slavery*, 150–51; Fraser, *Charleston!*, 228–29; J. Johnson, "Map of Charleston and its Defenses"; "Bird's Eye View" map. City residents believed that living on the Neck during the summer was "tantamount to ordering one's coffin."

24. Pratt, "Present," 229; "The Phosphate and Manufacturing Interests of South Carolina," *CDC*, 30 December 1868, 2; Chazal, *Century*, 62–63; "Cotton Planters are invited . . . ," *CDC*, 12 July 1869, 1; Mazyck, *Guide*, 166; Wines, *Fertilizer in America*, 120–22; "The Charleston Phosphates," *CDC*, 25 December 1868, 2; W. W. Memminger, "The Manufacture of Commercial Fertilizers," *RC* 4 (January 1873): 202–5.

25. *Census 1880 Manufactures, Charleston County*; "Etiwan Phosphate Co.," map; Land, *Charleston*, 137.

26. "Pacific Guano," *CDC*, 31 August 1869, 1; "Map of Charleston," (1883); Shepard Jr., "Report 1870," 4–5; Pacific Guano, *History*, 13.

27. "Works of the Wando Mining and Manufacturing Company," *RC* 2 (November 1870): 109–10; "There's Millions in It," Supplement to *CNC*, 1 March 1884, 1;

"The Phosphate Interests of Charleston," *CDC*, 9 February 1869, 2; Land, *Charleston*, 112; Chazal, *Century*, 63; "Map of Charleston," (1883); "Wando Phosphate Company (New Works)," map; Willis, "Phosphates," (1882).

28. Shepard Jr., "Report 1870," 4–5; "Atlantic Phosphate Works," *CDC*, 3 March 1871, 1; Mazyck, *Guide*, 168–73; Land, *Charleston*, 145–46; "Atlantic Phosphate Company," map; "Map of Charleston," (1883); Gibbes, "Report," 4–5.

29. "Report of the Royalty Due and Paid from September 1, 1878, to September 1, 1879," *Reports 1879*, 116; *Census 1880 Manufactures, Charleston County*; F. L. Frost, "The Stono Phosphate Company," *CNC*, 12 July n.y., EWS-7; "A Miracle in Milling . . . ," [*CNC* 4 May 1880?], EWS-9.

30. "Map of Charleston," (1883); Willis, "Marl Beds," 68; Land, *Charleston*, 173, 179; *Ashepoo Prospectus*, 3–6; Chazal, *Century*, 67; Fairly v. Wappoo Mills (1894), in Shand, *Reports of Cases*, 229; "Wappoo Mills Phosphate Co.," map; "Ashepoo Fertilizer Company," map. In the late 1870s Sardy sold Wappoo to C. C. Pinckney Jr. and the Ashepoo plant to Robertson, Taylor & Co.

31. "Table-Fertilizers Georgia 1877–78"; T. Taylor, "Statement" (1879), 121; "For the Season of 1880–81, by the Department of Agriculture of South Carolina," EWS-6; Willis, "Phosphates," (1883); "South Carolina Phosphates," (1883); "Map of Charleston," (1883); Coulter, *Williams*, 139, 153–54, 162–63; Wines, *Fertilizer in America*, 146.

32. "The Phosphate and Manufacturing Interests of South Carolina," *CDC*, 30 December 1868, 2; "Pacific Guano," *CDC*, 31 August 1869, 1; "Atlantic Phosphate Works," *CDC*, 3 March 1871, 1.

33. Markham, *Fertilizer Industry*, 63; "South Carolina—The South in 1880," 56; "The Phosphate Interests of Charleston," *CDC*, 9 February 1869, 2. Superphosphate manufacture had always involved "relatively simple technology" and "fairly modest capital outlays." Phosphate rock and sulfuric acid made up 95 percent of production costs (1950s), with mixing facilities, storage space, and labor making up the remaining 5 percent.

34. Memminger, "Manufacture," 204–5; "The Phosphate and Manufacturing Interests of South Carolina," *CDC*, 30 December 1868, 2; Wines, *Fertilizer in America*, 137–39. Workers produced one "batch" at a time instead of a continuous stream of product.

35. Pratt, "Present," 229; "The Phosphate and Manufacturing Interests of South Carolina," *CDC*, 30 December 1868, 2; Chazal, *Century*, 62–64. Fearing competition, northern sulfuric-acid producers refused Pratt access to their works. He claimed that SASP raised the level of "soluble bone phosphate" from 8–10 percent to 26–33 percent and "compelled the Northern manufacturer to do likewise or lose his trade."

36. "Etiwan Guanos," front matter; SASP, *Almanac 1872*, front matter; "Queries on Composting Commercial Fertilizers," *RC* 3 (January 1872): 184–86; F. Holmes, *Phosphate Rocks*, 78, ads.

37. "The Charleston Phosphates," *RC* 1 (October 1869): 49; John S. Reese & Co.

(Baltimore) to J. N. Robson (Charleston), 15 January 1871, John N. Robson Papers; "Circular of John S. Reese & Co.," 175–76.

38. Stono Phosphate Company, "Reply of President and Board," 1–8, 11. Agent-treasurers hired subagents, ordered materials, controlled production, made sales contracts, marketed the product, established sales commissions, and kept the books. Shareholders, not agents, sustained the losses from late or abandoned credit sales.

39. Stono Phosphate Company, "Reply of President and Board," 6–12; Mazyck, *Guide*, 173; "Copartnership Directory," *Charleston Directory 1872–73*, 290.

40. Ayers, *Promise of the New South*, 83–87.

41. "An Act to Amend the Charter of the Sulphuric Acid and Superphosphate Company," *Acts 1877–78*, 330; Mazyck, *Guide*, 167; "The Sulphuric Acid and Phosphate Company's Works," *RC* 2 (November 1870): 108–9; Wines, *Fertilizer in America*, 133–34.

42. "The Pacific Guano Company," *CDC*, 26 October 1869, 1; SASP, *Almanac 1872*, 1–20; Wines, *Fertilizer in America*, 124; "Fertilizers," *CDC*, 1 November 1872, 4; Sardy, *Sardy's Almanac 1872*, 1–16. Acid Phosphate had to be composted with cottonseed.

43. "Fertilizers," *CDC*, 1 November 1872, 4; Carolina Fertilizer ad, *CDC*, 25 February 1869, 3; Williams & Co., "Carolina Almanac 1873," 1–18; Coulter, *Williams*, 153–54.

44. Stono, *Almanac 1871*, 1, 16–19; Stono, *Almanac 1872*, 18–23; Stono, *Almanac 1880*, 1; Carolina Fertilizer ad, *CDC*, 25 February 1869, 3; Williams & Co., "Carolina Almanac 1873," 1–18.

45. "Soluble Pacific Guano," SCHS.

46. Stono, *Almanac* 1880, 2–4.

47. Stono, *Almanac 1871*, 19; Stono, *Almanac 1872*, 20.

48. "Cotton Planters are invited . . . ," *CDC*, 12 July 1869, 1; SASP, *Almanac 1872*, cover, 42–51; G. Wright, *Old South, New South*, 30–35; Wines, *Fertilizer in America*, 124, 148, 156, 167–70.

49. "Queries on Composting," 184–86; Memminger, "Manufacture," 202–5.

50. *Census 1870 Industry, Charleston County*; *Census 1870 Population, Charleston County*; "The Labor Question," *CNC*, 12 September 1873, 4; "The Strike," *CNC*, 15 September 1873, 4; Powers, *Black Charlestonians*, 131–33. SASP's pre-strike daily wages are unknown.

51. "Mob Law in the City," *CNC*, 11 September 1873, 4; "The Phosphate Mills Strikers," *CNC*, 15 September 1873, 4; "The Strikers Again," *CNC*, 20 September 1873, 4; Powers, *Black Charlestonians*, 133.

52. "The Labor Question," *CNC*, 10 September 1873, 4; "Mob Law in the City," *CNC*, 11 September 1873, 4.

53. "A Strange Proceeding," *CDC*, 2 October 1869, 1; "The Longshoremen's Strike," *CDC*, 6 October 1869, 1; Powers, *Black Charlestonians*, 126–30; Arnesen, *Waterfront Workers*, 21–25, 32–33, 54; Kolchin, *First Freedom*, 130.

54. *Census 1870 Manufactures, Charleston County*; *Census 1880 Manufactures, Charleston County*; Carlton, *Mill and Town*, 85–86n7.

55. Robert B. S. Sanders, plaintiff, appellant, v. The Etiwan Phosphate Company, defendant, respondent, Argument submitted for respondent by AT Smythe, S.C. Supreme Court, First Circuit (1883), 1–6, 22, SCHS.

56. "Bird's Eye View" map; James P. Allen, "Sub Division of the Allen Farm on Charleston Neck" (May 1878), SCHS; "Map of Charleston," (1883); Gibbes, "Report," 3–5, 8; Memminger, "Manufacture," 205.

57. Mazyck, *Guide*, 171; *Census 1880 Population, Charleston County*; "Map of Charleston," (1883). The average age of the seventy-one black men was thirty-two years.

58. "Map of Charleston," (1883); "South Carolina Agricultural Society-Meeting of the Executive Committee," *CDC*, 10 February 1873, 1; John Kennerty v. The Etiwan Phosphate Company, Copy Summons-for Relief, Charleston County Court of Common Pleas (1881), 6–8, 18, 28, SCHS; "Sulphurous Acid Fumes and Vegetation," *RC* 4 (September 1873): 632–34; "Fumes of Phosphate Works: A Report from a Committee of the Agricultural Society of South Carolina that their Injurious Effect on Surrounding Vegetation is Almost Inappreciable" (1879), SCHS.

59. Kennerty v. Etiwan, Common Pleas (1881), 3, 8–22, 25–31; "Industrial and Commercial Charleston: Map"; Dewey, "The Fickle Finger of Phosphate," 572–74. In the 1950s, researchers in Florida discovered that processing phosphate rock with sulfuric acid released fluorides that were dangerous to humans and especially harmful to truck crops.

60. Goode, *History of the Menhaden*, 490; "An Act to Prohibit the Digging, Mining, or Removing of Phosphate Rocks . . . without License" and "Joint Resolution . . . Define and Protect the Interest of the State Therein," *Acts Extra Session, 24 April 1877–9 June 1878*, 305, 320–21; "Phosphates," *Reports 1879*, 103–4, 109; "Phosphate Mining," *Reports 1880*, 110–11; Roche, "Phosphate Department," (1881), 14–15, 18, EWS-6; "Phosphate Litigation," *Reports 1886 v. 1*, 135–36.

61. Robert Adger et al v. Farmers' Phosphate Company, S.C. Supreme Court, First Cir., Charleston County, Statement of Case and Exceptions on Appeal (Charleston: News and Courier Book Presses, 1879), 9–10, 21, MSP; G. C. Rogers, *Lawyers*, 80–82; E. L. Roche, "Report of the Special Assistant," *Reports 1882*, 289–92; "Phosphate Department," *Reports 1883*, 928–29, 949–50; State of South Carolina v. Pacific Guano Company, Record, Beaufort County Court of Common Pleas (Charleston: News and Courier Book Presses, 1883), 3, 345–46, MSP; "Phosphate Litigation," *Reports 1886, v.1*, 136–37.

62. Day, "Fertilizers," 787; "Phosphate Litigation," *Reports 1886, v.1*, 137–38; Land, *Charleston*, 152; E. L. Roche, "Report of the Special Assistant," *Reports 1884, v.2*, 647–48; "The Phosphate Department," and "Report of Special Assistant Phosphate Department," *Reports 1889 v.2*, 713–14, 729–31; Gaines, "Pacific Guano Company," 15. Gaines writes that the company's demise "is shrouded in mystery" and suggests that embezzlement played a role.

63. Wines, *Fertilizer in America*, 122–24, 157–59; Day, "Fertilizers," 825; Millar, *Florida, South Carolina*, 176.

64. Wines, *Fertilizer in America*, 110, 114, 123–24, 136–44, 222n7; Jacob, "History and Status," 48, 56.

65. "List of Prices: Fertilizers," *CDC*, 22 November 1867, 3 April 1868, 23 July 1869, 1 September 1869, 18 February 1870, 4 April 1873, 4; Wines, *Fertilizer in America*, 81–82, 123–24, 142, 157–59; C. Wright, *Sixth Special Report*, 103; G. Wright, *Old South, New South*, 30; Hahn, *Roots of Southern Populism*, 143–46; R. Taylor, "Southeast Part I," 311–12.

66. G. Wright, *Old South, New South*, 29–38, 45–49; Earle, "The Price of Precocity," 51–54.

67. "The Phosphate and Manufacturing Interests of South Carolina," *CDC*, 30 December 1868, 2; D. H. Jacques, "Commercial Fertilizers," *RC* 3 (July 1872): 543.

Conclusions and Epilogue

1. F. Holmes, *Phosphate Rocks*, 73.

2. Roark, *Masters Without Slaves*, 208–9.

3. Armes, *Story of*, 330–40, 424–28.

4. Jeff Forret and others have begun investigating these underground economies during the antebellum era. See Forret, *Race Relations* and "Slaves, Poor Whites, and the Underground."

5. Rabinowitz, *First New South*, 1–6; Woodward, *Origins*, ix, x.

6. Shick and Doyle, "Boom," 1–4, 24–31; D. Doyle, *New Men*, 80, 111, 117–29, 159–60, 174–75, 186–90, 225.

7. Fraser, *Charleston!*, 339, 385–88, 405–19; SC Department of Agriculture, *South Carolina: A Handbook*, 87; Arlie Porter, "EPA Evaluates Phosphate Plant Pollution," *CPC*, 10 October 1999, A1; Arlie Porter, "Macalloy Brought Jobs, Pollution," *CPC*, 6 November 2000, A1; Jack Leland, "Fertilizer Industry: Strike Brought About Police Force," *Charleston Evening Post*, 18 May 1970, 8-B; Shick and Doyle, "Boom," 1.

8. Holden, *Maelstrom*, 114; Chazal, *Century*, 56–61; Edgar, *South Carolina*, 451; Leland, "Fertilizer," 8-B; Malde, "Geology of the Charleston Phosphate Area," 72.

9. Jason Hardin, "Planners Envision Thriving Landscape for the Neck," *CPC*, 15 July 2002, A1; Chazal, *Century*, 62–68; Shick and Doyle, "Boom," 27–29.

10. Arlie Porter, "Legacy of Contamination Still Haunts Rivers, Creeks," *CPC*, 24 February 1998, A7; Porter, "EPA Evaluates Phosphate Plant Pollution," *CPC*, 10 October 1999, A1; Porter, "Macalloy Brought Jobs, Pollution," *CPC*, 6 November 2000, A1.

Bibliography

Primary Sources

Ackerman, Ann Lesesne. Conversation with author. Chisolm Island, South Carolina, 15 June 2001.

Adger, John B. *My Life and Times, 1810–1899*. Richmond: Presbyterian Committee of Publication, 1899.

Adger Family. Papers. South Carolina Historical Society, Charleston.

American Agriculturist 16 (Jan. 1857): 12.

Ansted, David T. *Geology, Introductory, Descriptive, and Practical*. London: John Van Voorst, 1844.

"Ashepoo Fertilizer Company." Map. New York: Sanborn Map Company, Insurance Maps, 1902.

Ashepoo Mining and Manufacturing Guano Co., *Prospectus*. New York: George F. Nesbitt, 1874, EWS-5.

Atlantic Phosphate Company. *Almanac, 1872*. Charleston: Walker, Evans & Cogswell, EWS-3.

"Atlantic Phosphate Company." Map. New York: Sanborn Map Company, Insurance Maps, 1884.

Baker and Turner. Letterbook. Middleton Place Archives, Charleston.

Bank of Charleston. Collection. South Carolina Historical Society, Charleston.

"Beaufort County." Map. Cincinnati: Mass Marketing, 2000.

Beaufort Gazette, 1992.

Beaufort Republican, 1872–73.

(Beaufort) *Tribune and Commercial*, 1878, EWS-7.

"B. H. Rutledge." *Transactions of the Huguenot Society of South Carolina* 3 (1894): 26–27.

"Bird's Eye View of the City of Charleston South Carolina." Map. N.p.: C. Drie, 1872.

Bowens, Richmond. Interview by author. Tape recording. Drayton Hall, Charleston, 21 July 1997.

Bradley, Julie Powers. E-mail to author, 2 February 2000, 26 March 2002.

Chapel Hill Iron Mountain Company. Records. Southern Historical Collection, University of North Carolina, Chapel Hill, N.C.

Charleston and Vicinity Illustrated: South Carolina Inter-State and West Indian Exposition. Charleston: Walker, Evans & Cogswell, 1901.

Charleston City Directory for 1872–1873. Charleston: Walker, Evans & Cogswell, 1872.

Charleston Daily Courier, 1867–73.

Charleston Evening Post, 1970.

Charleston Mercury, 1868.

Charleston News and Courier, 1873–1910.

Charleston Post and Courier, 1995–2002.

Charleston Register of Mesne Conveyance 1800–1881.

Charleston, South Carolina, Mining and Manufacturing Company. Minutes of Meetings Stockholders and Directors, 1927–38. Mobil Corporation Corporate Archives. Fairfax, Virginia.

Charlotte Observer, 1998.

Chicora Mining & Manufacturing Company 1870–72. Records. Perkins Library Special Collections, Duke University, Durham, N.C.

City of Charleston and the State of South Carolina. N.p., c. 1889, Charleston Museum.

City of North Charleston Historical and Architectural Survey: Final Survey Report. Charleston: Preservation Consultants, 1995.

Conner, James. "Report of the Attorney General to the General Assembly of South Carolina, Concerning the Phosphate Interests of the State, under Joint Resolution approved June 9, 1877." Columbia: Calvo & Patton, State Printers, 1878.

Coosaw Company. Records. South Carolina Historical Society, Charleston.

CRM. "How Fertilizers are made in Atlanta." *Scientific American Supplement* 316 (21 Jan. 1882): 5041.

Day, David T. "Fertilizers: Phosphate Rock" and "Manufactured Fertilizers." In *Mineral Resources of the United States 1883–1884,* United States Geological Survey, 783–88, 815–26. Washington, D.C.: Government Printing Office, 1885.

"Death of Dr. N. A. Pratt." *The American Fertilizer* 25, no. 5 (1906): 16.

Doyle, Barbara. Letter to author, 23 February 2001, 14 September 2001.

Drayton, John. Papers. South Carolina Historical Society, Charleston.

Drayton Hall Historic Structures Report, Vol. 1, Drayton Hall Archives, Charleston.

"Edisto Nature Trail." Map 8. N.p.: Westvaco Corporation, Westvaco Timberlands Division, n.d.

Edisto Phosphate Company. Records. South Carolina Historical Society, Charleston.

"Etiwan Guanos, or Soluble Phosphate Manures of the Sulphuric Acid and Superphosphate Company, Charleston, S.C." Charleston: Walker, Evans & Cogswell, 1870. http://digital.tcl.sc.edu/cdm/compoundobject/collection/phosphate/id/656/rec/9.

Etiwan Phosphate Company. Records. South Carolina Historical Society, Charleston.

"Etiwan Phosphate Company" Map. New York: Sanborn Map Company, Insurance Maps, 1884.

Freedley, Edwin T. *Philadelphia and Its Manufacturers*. Philadelphia: Edward Young, 1867.

George W. Williams & Co. "The 'Carolina Fertilizer' Almanac and Farmers' Journal for 1873." Supplement to *RC*. Charleston: Walker, Evans & Cogswell, 1873, EWS-5.

Gibbes, James S. "Report of James S. Gibbes, Esq., President [pro tem] of the Stono Phosphate Company, of Charleston, as read before the annual meeting of the stockholders of the company 3d May 1871, and published by order of the Board of Directors." Charleston: Walker, Evans & Cogswell, 1871, EWS-3.

Hammond, Harry. *South Carolina: Resources and Population, Institutions and Industries*. Charleston: Walker, Evans & Cogswell, 1883.

Hemphill, J. C., ed. *Men of Mark in South Carolina*, III. Washington: Men of Mark Publishing, 1908.

Hinson Collection. Charleston Library Society, Charleston.

Hoke, Robert F. Papers. Southern Historical Collection, University of North Carolina, Chapel Hill, N.C.

Holmes, Francis S. *Phosphate Rocks of South Carolina and the Great Carolina Marl Bed*. Charleston: Holmes' Book House, 1870.

"Industrial and Commercial Charleston: Map of Charleston, S.C. and Vicinity." Map. Charleston: McCrady Bros. & Cheves, 1915, Charleston County Public Library.

Johnson, John. "Map of Charleston and its Defenses," 1863. Middleton Place Archives, Charleston.

Jowitt, Thad. C. *Jowitt's Illustrated Charleston City Directory and Business Directory 1869–70*. Charleston: Walker, Evans, & Cogswell, 1870.

Judd, William R. "A Report on the Ruins of the Dennis, Washington and Nanny Notes Housesites, Drayton Hall: A Visual Survey February 1998." Drayton Hall Archives, Charleston.

———. "The Roberts/McKeever Housesite Ruins. Drayton Hall. A Visual Survey Nov./Dec. 1997." Drayton Hall Archives, Charleston.

Kennerty, Thomas J., and John Kennerty. Papers. South Carolina Historical Society, Charleston.

Kerr, Washington Caruthers. "Geological Map of North Carolina." Southern Historical Collection, University of North Carolina, Chapel Hill, N.C.

———. "North Carolina As a Place for Investment, Manufactures, Mining, Stock Raising, Fruit and Farming: What Northern Residents in North Carolina Say of It As a Place to Live In." Raleigh: The Observer, State Printer & Binder, 1879.

———. "Report on the Cotton Production of the State of North Carolina, With a

Discussion of the General Agricultural Features of the State." Washington, D.C.: U.S. Government Printing Office, 1884.

Land, John E. *Charleston: Her Trade, Commerce and Industries, 1883–4*. Charleston: JE Land, 1884.

Marine and River Phosphate Company, Charleston, S.C. February 26 1883. Charleston: Walker, Evans & Cogswell, 1883.

Marine and River Phosphate Mining & Manufacturing Company, of South Carolina, *By-Laws*. Charleston: Walker, Evans & Cogswell, 1870, EWS-3.

Mazyck, Arthur. *Guide to Charleston Illustrated*. Charleston: Walker, Evans, & Cogswell, 1875.

McCrady, Edward Jr., "The Phosphate Question Discussed and the Arguments of the Opponents of the Act of March 22, 1878 Replied To." College of Charleston Special Collections, Charleston.

Middleton, Williams. Letterbook 1868–70. Transcribed by Barbara Doyle. Middleton Place Archives, Charleston.

Middleton, Williams. Papers. Middleton Place Archives, Charleston.

Millar, C. C. Hoyer. *Florida, South Carolina and Canadian Phosphates*. London: Eden Fisher, 1892.

Mitchell & Smith. Papers. South Carolina Historical Society, Charleston.

Morfit, Campbell. *A Practical Treatise on Pure Fertilizers; and the Chemical Conversion of Rock Guanos, Marlstones, Coprolites, and the Crude Phosphates of Lime and Alumina Generally, into Various Products*. New York: D. Van Nostrand, 1872.

Moses, Otto A. "Phosphate Deposits of South Carolina." In *Mineral Resources of the United States 1882*. United States Geological Survey, 504–21. Washington, D.C.: Government Printing Office, 1883.

National Register of Historic Places Inventory, 1874. Charleston City Archives, Charleston.

Nepveux, Ethel Trenholm Seabrook. E-mail to author, 17 August 2001, 6 February 2003.

New York Times, 1868, 1881.

Northen, W. F., ed. *Men of Mark in Georgia*. Vol. 5. Atlanta: A. B. Caldwell, 1908.

"Pacific Guano Company." Map. New York: Sanborn Map Company, Insurance Maps, 1884.

Pacific Guano Company. *The Pacific Guano Company: Its History; Its Products and Trade: Its Relation to Agriculture*. Cambridge: Riverside Press, 1876.

Pamphlets. South Carolina Historical Society, Charleston.

Payne, Robert K. "Plan of the Wappoo Mills Tract of Land." 1857, 1899. Charleston City Archives.

Pelzer Manufacturing Company, *Charter and By-Laws*. Charleston: Walker, Evans & Cogswell, 1881.

Penn School. Papers. Penn Center, Frogmore, St. Helena Island, South Carolina.

Phosphates Collection. South Carolina Historical Society, Charleston.

Port Royal Commercial, 1873.

Pratt, Nathaniel A. *Ashley River Phosphates. History of the Marls of South Carolina, and of the Discovery and Development of the Native Bone Phosphates of the Charleston Basin*. Philadelphia: Inquirer Book & Job Print, 1868, EWS-1.

———. "Southern Phosphates—Their Past, 1840 to 1867." [*Dixie* 1892?]: 146–49, "Notes, clippings, etc. on Phosphates by Maj. E.A. Willis, and other phosphate material," Charleston Museum Archives.

———. "Southern Phosphates—Their Present." [*Dixie* 1892?]: 224–30, "Notes, clippings, etc. on Phosphates by Maj. E.A. Willis, and other phosphate material," Charleston Museum Archives.

Pratt, Nathaniel A. Papers. Perkins Library Special Collections, Duke University, Durham, N.C.

"Quarrying Phosphates in South Carolina." *American Agriculturist* 31 (Jan. 1872): 20.

Ravenel, William. "Reply of William Ravenel, Esq., President of the Stono Phosphate Company, of Charleston, So. Ca. to Arthur M. Huger, Esq., late Secretary and Treasurer." Charleston: Walker, Evans & Cogswell, 1878.

Raymond, Rossiter W. "Historical Sketch of Mining Law." In *Mineral Resources of the United States 1883–1884*, United States Geological Survey, 988–1004. Washington, D.C.: Government Printing Office, 1885.

Reese, John S. and Company. "Circular of John S. Reese & Co." *DeBow's Review* 6 (Feb. 1869): 175–77.

"Retirement of Col. George W. Scott." *The American Fertilizer* 9 (1898): 120.

Robson, John N. Papers. Perkins Library Special Collections, Duke University, Durham, N.C.

Rogers, G. Sherburne. "The Phosphate Deposits of South Carolina." *U.S. Geological Survey Bulletin* 580 (1914): 183–220.

Ruffin, Edmund. *Report of the Commencement and Progress of the Agricultural Survey of South Carolina for 1843*. Columbia: A.H. Pemberton, 1843.

Rural Carolinian, 1869–73.

Sardy, John B. *Sardy's Phosphate Almanac 1872*. Charleston: Walker, Evans, & Cogswell, 1872, EWS-3.

Shaler, N. S. *On the Phosphate Beds of South Carolina*. Boston: A.A. Kingman, 1870.

Shand, Robert W. *Reports of Cases Heard and Determined by the Supreme Court of South Carolina Volume XLIV*. Columbia: R.L. Bryan, 1896. http://books.google.com/books?id=4rwaAAAAYAAJ&printsec=frontcover&source=gbs_ge_summary_r&cad=0#v=onepage&q&f=false.

Shepard, Charles U., Jr. "Report of Charles U. Shepard Jr., Inspector of Guano and Fertilizers, to the General Assembly of South Carolina, November 1870." Columbia: J.W. Denny, 1871. EWS-2.

———. "South Carolina Phosphates. A Lecture Delivered Before the Agricultural Society of South Carolina, Charleston, South Carolina, December 12, 1879."

Charleston: News and Courier Book Presses, 1880. EWS-"SC Phosphates unsorted."

Sholes' Directory of the City of Charleston 1881–82. Charleston: Sholes, 1881, 1882.

Simons and Locke Law Firm. Records. South Carolina Historical Society, Charleston.

Smith, Alice R. H. "One or two hands in the barn-yard." Sketch in *A Woman Rice Planter* by Elizabeth W. Allston Pringle, with an introduction by Owen Wister and illustrations by Alice R. H. Smith. New York, Macmillan, 1913.

Smith, H.A.M. Map. *South Carolina Historical and Genealogical Magazine* 19 (Jan. 1918): 1.

"Soluble Pacific Guano and Compound Acid Phosphate." Charleston: Walker, Evans, & Cogswell, 1877, Pamphlet, SCHS.

South Carolina Department of Agriculture, Commerce, and Industries and Clemson College. *South Carolina: A Handbook*. Columbia: Department of Agriculture, Commerce, and Industries, 1927.

South Carolina General Assembly. *Acts and Joint Resolutions of the General Assembly of the State of South Carolina*. Columbia: various, 1868–83.

———. *Reports and Resolutions of the General Assembly of the State of South Carolina*. Columbia: various, 1871–1901.

South Carolina House of Representatives. *Journal of the House of Representatives of the State of South Carolina*. Columbia: various, 1869–71.

South Carolina in 1884: A View of the Industrial Life of the State. Charleston: News & Courier Book Presses, 1884.

South Carolina Institute Premium List. Fair of 1870. Charleston: Walker, Evans & Cogswell, 1870.

South Carolina Phosphate and Phosphatic River Mining Co. *By-Laws*. Charleston: Walker, Evans & Cogswell, 1871.

South Carolina State Senate. *Journal of the Senate of the State of South Carolina*. Columbia: various, 1869–70.

"South Carolina—The South in 1880—A Glimpse of the Industrial Interests of Charleston." *Frank Leslie's Illustrated Newspaper* (Sept. 25, 1880): 56. http://digital.tcl.sc.edu/cdm/singleitem/collection/phosphate/id/728/rec/2.

Stono Phosphate Company. *Almanac and Hand Book, 1871*. Charleston: Walker, Evans & Cogswell, 1871, Pamphlet, SCHS.

———. *Almanac and Hand Book, 1872*. Charleston: Walker, Evans & Cogswell, 1872, EWS-3.

———. *Almanac and Hand Book, 1880*. Charleston: Lucas and Richardson, 1880, Pamphlet, SCHS.

———. "Reply of the President and Board of Directors of the Stono Phosphate Company of Charleston, So. Ca. to Messrs. John H. Devereux and Paul C. Trenholm, Committee, &c." Charleston: Walker, Evans & Cogswell, 1878.

———. *Revised By-Laws*. Charleston: Walker, Evans & Cogswell, 1879.

"Stono Phosphate Company." Map. New York: Sanborn Map Company, Insurance Maps, 1884.

Sulphuric Acid & Superphosphate Co. *Almanac 1872*. Charleston: Walker, Evans & Cogswell, 1871, EWS-3.

Taylor, Frank E. Papers. Perkins Library Special Collections, Duke University, Durham.

Taylor, Frank E. Papers. South Caroliniana Library, University of South Carolina, Columbia.

Tuomey, Michael. *Report on the Geological and Agricultural Survey of the State of South Carolina: 1844*. Columbia: A.S. Johnston, 1844.

———. *Report on the Geology of South Carolina*. Columbia: A.S. Johnston, 1848.

Tuomey, Michael, and Francis S. Holmes. *Pleiocene Fossils of South Carolina: Containing Descriptions and Figures of the Polyparia, Echinodermata and Mollusca*. Charleston: Russell and Jones, 1857.

U.S. Bureau of the Census. *Ninth Census of the United States, 1870: Compendium of the Ninth Census*. Washington, D.C.: U.S. Government Printing Office, 1872.

———. *Ninth Census of the United States, 1870: Industry and Wealth*. Vol. 3. Washington, D.C.: U.S. Government Printing Office, 1872.

———. *Ninth Census of the United States, 1870: Manuscript Schedule of Industry 1870, South Carolina, Beaufort, Charleston, and Colleton Counties.*

———. *Ninth Census of the United States, 1870: Manuscript Schedule of Population 1870, South Carolina, Beaufort, Charleston, and Colleton Counties.*

———. *Ninth Census of the United States, 1870: Statistics of Population*. Vol. 1. Washington, D.C.: U.S. Government Printing Office, 1872.

———. *Tenth Census of the United States, 1880: Compendium of the Tenth Census*. Part 1. Washington, D.C.: U.S. Government Printing Office, 1883.

———. *Tenth Census of the United States, 1880: Manuscript Schedule of Manufactures, South Carolina, Beaufort, Charleston, and Colleton Counties.*

———. *Tenth Census of the United States, 1880: Manuscript Schedule of Population, South Carolina, Beaufort, Charleston, and Colleton Counties.*

U.S. Congress. *Testimony Taken by the Joint Select Committee to Inquire into the Condition of Affairs in the Late Insurrectionary States*. 13 vols. Washington, D.C.: Government Printing Office, 1872.

United States Geological Surveys. *Mineral Resources of the United States 1882–1891*. Washington, D.C.: Government Printing Office, 1883–92.

Waggaman, William Henry. "A Report on the Phosphate Fields of South Carolina." *U.S. Department of Agriculture Bulletin* 18 (1913): 1–12.

Wando Mining and Manufacturing Company. *Almanac 1872*. Charleston: Wm. C. Dukes, 1871, EWS-1.

"Wando Phosphate Company." Map. New York: Sanborn Map Company, Insurance Maps, 1884.

Wappoo Mills Phosphate Co." Map. New York: Sanborn Map Company, Insurance Maps, 1884.

Williams, George W. Papers. South Carolina Historical Society, Charleston.

Willis, Edward. Scrapbooks. Charleston Museum Archives, Charleston.

———. "The Marl Beds and Phosphate Rocks of South Carolina, A Statistical Statement made at the request of E. H. Frost, Esq., Chairman of the Statistics of the Charleston Chamber of Commerce." N.p., 1872. EWS-2.

Wilson's Business Directory of New York City 1869–70. New York: J.F. Trow, 1869.

Wright, Carroll D. *Sixth Special Report of the Commissioner of Labor: The Phosphate Industry of the United States*. Washington, D.C.: Government Printing Office, 1893.

Wyatt, Francis. *The Phosphates of America: Where and How They Occur; How They Are Mined; and What They Cost*. New York: Scientific Publishing, 1891.

Year Book—1883 City of Charleston, So. Ca. Charleston: News & Courier Book Presses, 1883.

Year Book—1886 City of Charleston, So. Ca. Charleston: Walker, Evans & Cogswell, 1886.

Secondary Sources

Abbott, Martin. *The Freedmen's Bureau in South Carolina, 1865–1872*. Chapel Hill: University of North Carolina Press, 1967.

Allen, Glover M. "Fossil Mammals from South Carolina," *Bulletin of the Museum of Comparative Zoology at Harvard College* 67 (Jul. 1926): 447–67.

Allmendinger, David F., Jr. *Ruffin: Family and Reform in the Old South*. New York: Oxford University Press, 1990.

———. "The Early Career of Edmund Ruffin, 1810–1840." *Virginia Magazine of History and Biography* 93 (Apr. 1985): 127–54.

———, ed. *Incidents of my Life: Edmund Ruffin's Autobiographical Essays*. Charlottesville: University Press of Virginia, 1990.

Alston, Lee J., and Kyle D. Kauffman. "Up, Down, and Off the Agricultural Ladder: New Evidence and Implications of Agricultural Mobility for Blacks in the Postbellum South." *Agricultural History* 72 (Spring 1998): 263–79.

Armes, Ethel. *The Story of Coal and Iron in Alabama*. 1910. Reprint, New York: Arno Press, 1973.

Armstrong, Thomas F. "From Task Labor to Free Labor: The Transition Along Georgia's Rice Coast, 1820–1880." *Georgia Historical Quarterly* 64 (Winter 1980): 432–47.

———. "The Christ Craze of 1889: A Millennial Response to Economic and Social Change." In *Toward a New South? Studies in Post-Civil War Southern Communities*, edited by O.V. Burton and R. C. McMath Jr., 223–45. Westport, Conn.: Greenwood Press, 1982.

Arnesen, Eric. "Up from Exclusion: Black and White Workers, Race, and the State of Labor History." *Reviews in American History* 26.1 (1998): 146–74.

———. *Waterfront Workers of New Orleans: Race, Class, and Politics, 1863–1923*. New York: Oxford University Press, 1991.

Ayers, Edward L. *The Promise of the New South: Life After Reconstruction*. New York: Oxford University Press, 1993.

Azrael, Louis. *The Story of Alex. Brown & Sons, 1800–1975*. Rev. ed. Baltimore: Alex. Brown & Sons, 1975.

Bailey, N. Louise, Mary L. Morgan, and Carolyn R. Taylor. *Biographical Directory of the South Carolina Senate 1776–1985*. Columbia: University of South Carolina Press, 1986.

Barefoot, Daniel W. *General Robert F. Hoke: Lee's Modest Warrior*. Winston Salem: John F. Blair Publisher, 1996.

Bateman, Fred, and Thomas Weiss. *A Deplorable Scarcity: The Failure of Industrialization in the Slave Economy*. Chapel Hill: University of North Carolina Press, 1981.

Beatty, Bess. *Alamance: The Holt Family and Industrialization in a North Carolina County, 1837–1900*. Baton Rouge: Louisiana State University, 1999.

Berlin, Ira. "Time, Space, and the Evolution of Afro-American Society on British Mainland North America." *American Historical Review* 85 (Feb. 1980): 44–78.

Billings, Dwight B., Jr. *Planters and the Making of a "New South": Class, Politics, and Development in North Carolina, 1865–1900*. Chapel Hill: University of North Carolina Press, 1979.

Blakey, Arch Fredric. *The Florida Phosphate Industry: A History of Development and Use of a Vital Mineral*. Cambridge: Harvard University Press, 1973.

Bulloch, James D. *The Secret Service of the Confederate States in Europe*. Vol. 1. 1883. Reprint, New York: Thomas Yoseloff, 1959.

Capers, Henry D. *The Life and Times of C. G. Memminger*. Richmond: Everett Waddey, 1893.

Carlton, David L. *Mill and Town in South Carolina, 1880–1920*. Baton Rouge: Louisiana State University Press, 1982.

Cecelski, David S. *The Waterman's Song: Slavery and Freedom in Maritime North Carolina*. Chapel Hill: University of North Carolina Press, 2001.

Chazal, Philip E. *The Century in Phosphates and Fertilizers, A Sketch of the South Carolina Phosphate Industry*. Charleston: Lucas-Richardson Lithograph & Printing, 1904.

Cheves, Langdon. "Middleton of South Carolina." *South Carolina Historical and Genealogical Magazine* 1 (Jul. 1900): 228–62.

Clifton, James M. "The Rice Driver: His Role in Slave Management." *South Carolina Historical Magazine* 82 (Oct. 1981): 331–53.

Cobb, James C. "Beyond Planters and Industrialists: A New Perspective on the New South." *Journal of Southern History* 54 (Feb. 1988): 45–68.

Coclanis, Peter A. "Entrepreneurship and the Economic History of the American South: The Case of Charleston and the South Carolina Low Country." In *Mar-*

keting in the Long Run: Proceedings of the Second Workshop on Historical Research in Marketing, edited by Stanley C. Hollander and Terence Nevett, 210–19. East Lansing: Board of Trustees, Michigan State University, 1985.

———. "How the Low Country Was Taken to Task: Slave-Labor Organization in Coastal South Carolina and Georgia." In *Slavery, Secession, and Southern History*, edited by Robert Louis Paquette and Louis A. Ferleger, 59–78. Charlottesville: University of Virginia Press, 2000.

———. "The Rise and Fall of the South Carolina Low Country: An Essay in Economic Interpretation." *Southern Studies* 24 (Summer 1985): 143–66.

———. *The Shadow of a Dream: Economic Life and Death in the South Carolina Low Country, 1670–1920*. New York: Oxford University Press, 1989.

Coker, P. C., III. *Charleston's Maritime Heritage, 1670–1865*. Charleston: CokerCraft Press, 1987.

Comfort, Jan. "Correspondence." *South Carolina Historical Magazine* 98 (Oct. 1997): 407.

Cooper, William J., Jr. *The Conservative Regime: South Carolina, 1877–1890*. Baltimore: Johns Hopkins University Press, 1968.

Cothran, James R. *Gardens of Historic Charleston*. Columbia: University of South Carolina Press, 1995.

Coulter, E. Merton. *George Walton Williams: The Life of a Southern Merchant and Banker 1820–1903*. Athens: Hibriten Press, 1976.

"David C. Ebaugh on the Building of 'The David.'" *South Carolina Historical Magazine* 54 (Jan. 1953): 32–36.

Davidson, Chalmers Gaston. *The Last Foray: The South Carolina Planters of 1860, A Sociological Study*. Columbia: University of South Carolina Press, 1971.

Degler, Carl N. "Rethinking Post-Civil War History." *Virginia Quarterly Review* 57 (Spring 1981): 250–67.

Dewey, Scott H. "The Fickle Finger of Phosphate: Central Florida Air Pollution and the Failure of Environmental Policy, 1957–1970." *Journal of Southern History* 65 (Aug. 1999): 565–603.

Donnelly, Ralph W. "Scientists of the Confederate Nitre and Mining Bureau." *Civil War History* 2 (Dec. 1956): 69–92.

Doster, James F. "Vicissitudes of the South Carolina Railroad, 1865–1878." *Business History Review* 30 (Jun. 1956): 175–95.

Downey, Tom. "Riparian Rights and Manufacturing in Antebellum South Carolina: William Gregg and the Origins of the 'Industrial Mind.'" *Journal of Southern History* 65 (Feb. 1999): 77–108.

Doyle, Don H. *New Men, New Cities, New South: Atlanta, Nashville, Charleston, and Mobile, 1860–1910*. Chapel Hill: University of North Carolina Press, 1990.

"Dr. N. A. Pratt, Scientist and Builder." *Commercial Fertilizer* 21 (Nov. 1920): 55–56.

Drago, Edmund L. *Initiative, Paternalism, and Race Relations: Charleston's Avery Normal Institute*. Athens: University of Georgia Press, 1990.

Dusinberre, William. *Them Dark Days: Slavery in the American Rice Swamps*. New York: Oxford University Press, 1996.

Earle, Carville. "The Price of Precocity: Technical Choice and Ecological Constraint in the Cotton South, 1840–1890." *Agricultural History* 66 (Summer 1992): 25–60.

Edgar, Walter. *South Carolina: A History*. Columbia: University of South Carolina Press, 1998.

Egerton, Douglas R. "Markets Without a Market Revolution: Southern Planters and Capitalism." *Journal of the Early Republic* 16 (Summer 1996): 207–21.

Engerman, Stanley L. "The Economic Response to Emancipation and Some Economic Aspects of the Meaning of Freedom." In *The Meaning of Freedom: Economics, Politics, and Culture After Slavery*, edited by Frank McGlynn and Seymour Drescher, 49–68. Pittsburgh: University of Pittsburgh Press, 1992.

Espenshade, Christopher T., and Marian D. Roberts. *An Archaeological and Historical Overview of the Drayton Hall Tract, Incorporating Data from the 1990 Archaeological Reconnaissance Survey*. Atlanta: Brockington & Associates, 1991.

Evans, Curtis. "Curt Evans Responds." Online posting. October 2002. H-NET Book Review. http://www.h-south@h-net.msu.edu.

Faunt, J.S.R., and R. E. Rector. *Biographical Directory of the South Carolina House of Representatives* Vol. 1. Columbia: University of South Carolina Press, 1974.

Faust, Drew Gilpin. *James Henry Hammond and the Old South: A Design for Mastery*. Baton Rouge: Louisiana State University Press, 1982.

Fenhagen, Mary Pringle. "Descendants of Judge Robert Pringle." *South Carolina Historical Magazine* 101 (Oct. 2000): 292–318.

Fishburne, Henry Gordon. *The Ladson Family of South Carolina and Georgia 1678–1900*. Spartanburg: The Reprint Company, 1995.

Flamming, Douglas. *Creating the Modern South: Millhands and Managers in Dalton, Georgia, 1884–1984*. Chapel Hill: University of North Carolina Press, 1992.

Fleetwood, Rusty. *Tidecraft: An Introductory Look at the Boats of Lower South Carolina, Georgia, and Northeastern Florida, 1650–1950*. Savannah: Coastal Heritage Society, 1982.

Fogel, Robert William, and Stanley L. Engerman. *Time on the Cross: The Economics of American Negro Slavery*. Boston: Little, Brown, 1974.

Foner, Eric. *Freedom's Lawmakers: A Directory of Black Officeholders during Reconstruction*. New York: Oxford University Press, 1993.

———. *Nothing But Freedom: Emancipation and Its Legacy*. Baton Rouge: Louisiana State University Press, 1983.

———. *Reconstruction: America's Unfinished Revolution 1863–1877*. New York: Harper & Row, Publishers, 1988.

Forret, Jeff. *Race Relations at the Margins: Slaves and Poor Whites in the Antebellum Southern Countryside*. Baton Rouge: Louisiana State University Press, 2006.

———. "Slaves, Poor Whites, and the Underground Economy of the Rural Carolinas." *Journal of Southern History* 60 (Nov., 2004): 783–824.

Fraser, Walter J., Jr. *Charleston! Charleston! The History of a Southern City*. Columbia: University of South Carolina Press, 1989.

Fuller, R. B. "The History and Development of the Mining of Phosphate Rock in the United States." In *Manual on Phosphates in Agriculture*, edited by Vincent Sauchelli, 29–37. Baltimore: Horn-Shafer, 1951.

Gagnon, Michael. Review of *The Old South's Modern Worlds: Slavery, Region, and Nation in the Age of Progress*, by L. Diane Barnes, Brian Schoen, and Frank Towers, eds. H-Southern-Industry. March, 2012. H-Net Reviews. http://www.h-net.org/reviews/showrev.php?id=34684.

Gaines, Jennifer Stone. "Pacific Guano Company." *Spritsail* 21 (Summer, 2007): 11–15. http://www.woodsholemuseum.org/spritsail/pacific_guano.pdf.

Garlington, J. C. *Men of the Time: Sketches of Living Notables*. 1902. Reprint, Spartanburg: The Reprint Company, 1972.

Geisst, Charles. *Monopolies in America: Empire Builders and their Enemies from Jay Gould to Bill Gates*. New York: Oxford University Press, 2000.

Genovese, Eugene D. *Roll, Jordan, Roll: The World the Slaves Made*. New York: Vintage Books, 1974.

———. *The Political Economy of Slavery: Studies in the Economy and Society of the Slave South*. New York: Vintage Books, 1967.

———. *The World the Slaveholders Made: Two Essays in Interpretation*. 1969. Reprint, Middletown, Conn.: Wesleyan University Press, 1988.

Glover, Beulah. *Narratives of Colleton County*. Brunswick, Ga.: Glover Printing, 1969.

Goode, G. Browne. *A History of the Menhaden*. New York: Orange Judd, 1880.

Griswold, Francis. *A Sea Island Lady*. New York: William Morrow, 1939.

Guerard, A. R. *A Sketch of the History, Origin and Development of the South Carolina Phosphates*. Charleston: Walker, Evans & Cogswell, 1884.

Hackney, Sheldon. "*Origins of the New South* in Retrospect." *Journal of Southern History* 38 (May 1972): 191–216.

Hagy, James W. *This Happy Land: The Jews of Colonial and Antebellum Charleston*. Tuscaloosa: University of Alabama Press, 1993.

Hahn, Steven. *The Roots of Southern Populism: Yeomen Farmers and the Transformation of the Georgia Upcountry, 1850–1890*. New York: Oxford University Press, 1983.

Hamrick, Tom. "To Sink a Yankee Ship. . . ." *South Carolina History Illustrated* 1 (Nov. 1970): 23–30.

Haskell, Helen W. *The Middleton Place Privy House: An Archaeological View of Nineteenth-Century Plantation Life*. Columbia: University of South Carolina Institute of Archaeology and Anthropology Popular Series 1, 1981.

Heyward, DuBose. *Mamba's Daughters: A Novel of Charleston*. 1929. Reprint, Columbia: University of South Carolina Press, 1995.

Hine, William C. "Black Politicians in Reconstruction Charleston, South Carolina: A Collective Study." *Journal of Southern History* 49 (Nov. 1983): 111–41.

———. "Dr. Benjamin A. Boseman Jr.: Charleston's Black Physician-Politician." In

Southern Black Leaders of the Reconstruction Era, edited by Howard N. Rabinowitz, 335–62. Urbana: University of Illinois Press, 1982.

Holden, Charles J. *In the Great Maelstrom: Conservatives in Post-Civil War South Carolina*. Columbia: University of South Carolina Press, 2002.

———. "'The Public Business is Ours': Edward McCrady, Jr. and Conservative Thought in Post-Civil War South Carolina, 1865–1900." *South Carolina Historical Magazine* 100 (Apr. 1999): 124–42.

Holman, Harriet R., ed. "Charleston in the Summer of 1841: the Letters of Harriott Horry Rutledge." *South Carolina Historical Magazine* 46 (Jan. 1945): 1–14.

Holmes, Henry Schulz. "The Trenholm Family." *South Carolina Historical and Genealogical Magazine* 16 (Oct. 1915): 150–60.

Holt, Thomas. *Black Over White: Negro Political Leadership in South Carolina during Reconstruction*. Urbana: University of Illinois Press, 1979.

Howe, J. J., and A. S. Howard. *Fossil Locations in South Carolina*. Museum Bulletin 3. Columbia: South Carolina Museum Commission, 1978.

Hudson, Larry E., Jr. Review of *Slave Counterpoint*, by Philip D. Morgan. *Journal of Southern History* 66 (May 2000): 381–83.

———. *To Have and to Hold: Slave Work and Family Life in Antebellum South Carolina*. Athens: University of Georgia Press, 1997.

Irwin, James R., and Anthony P. O'Brien. "Where Have All the Sharecroppers Gone? Black Occupations in Postbellum Mississippi." *Agricultural History* 72 (Spring 1998): 280–97.

Iseley, N. Jane, William P. Baldwin Jr., and Agnes L. Baldwin. *Plantations of the Low Country: South Carolina 1697–1865*. Greensboro: Legacy Publications, 1987.

Jacob, K. D. "History and Status of the Superphosphate Industry." In *Superphosphate: Its History, Chemistry, and Manufacture*, U.S. Department of Agriculture and Tennessee Valley Authority, 17–94. Washington, D.C.: Government Printing Office, 1964.

———. "Introduction, Scope, and Definitions." In *Superphosphate: Its History, Chemistry, and Manufacture*, U.S. Department of Agriculture and Tennessee Valley Authority, 1–7. Washington, D.C.: Government Printing Office, 1964.

Jacobs, W. P. *The Pioneer*. Clinton, S.C.: Jacobs Press, 1934.

Jakes, John. *Heaven and Hell*. New York: Signet, 2000.

Jenkins, Wilbert Lee. *Seizing the New Day: African Americans in Post-Civil War Charleston*. Bloomington: Indiana University Press, 1998.

Johnson, H. S., Jr. "Background and History of the 'South Carolina Geological Survey.'" *Geologic Notes* 3, no. 5 (Sept.–Oct. 1959): 5–6.

Johnson, Michael P. "Planters and Patriarchy: Charleston, 1800–1860." *Journal of Southern History* 46 (Feb. 1980): 45–72.

Johnson, Michael P., and James L. Roark, eds. *No Chariot Let Down: Charleston's Free People of Color On the Eve of the Civil War*. Chapel Hill: University of North Carolina Press, 1984.

Jordan, Laylon Wayne. "Between Two Worlds: Christopher G. Memminger and the Old South in Mid-Passage, 1830–1861." *Proceedings of the South Carolina Historical Association* (1981): 56–76.

Jordan, Weymouth T. "The Peruvian Guano Gospel in the Old South." *Agricultural History* 24 (Oct. 1950): 211–21.

Joyner, Charles. *Down by the Riverside: A South Carolina Slave Community*. Urbana: University of Illinois Press, 1984.

Kaye, Anthony E. "The Second Slavery: Modernity in the Nineteenth-Century South and the Atlantic World," *Journal of Southern History* 75 (Aug. 2009): 627–50.

Kelley, Robin D. G. "'We Are Not What We Seem': Rethinking Black Working-Class Opposition in the Jim-Crow South." *Journal of American History* 80 (Jun. 1993): 75–112.

Kent, Frank R. *The Story of Alex. Brown & Sons, 1800–1950*. Baltimore: Alex. Brown & Sons, 1950.

Kilbride, Daniel. "Southern Medical Students in Philadelphia, 1800–1861: Science and Sociability in the Republic of Medicine." *Journal of Southern History* 66 (Nov. 1999): 697–732.

Killick, John R. "The Cotton Operations of Alexander Brown & Sons in the Deep South, 1820–80." *Journal of Southern History* 43 (May 1977): 169–94.

King, Edward. *The Great South*. 1875. Reprint, Baton Rouge: Louisiana State University Press, 1972.

Kiser, Clyde Vernon. *Sea Island to City: A Study of St. Helena Islanders in Harlem and Other Urban Centers*. New York: Atheneum, 1969.

Kolchin, Peter. *American Slavery 1619–1877*. New York: Hill & Wang, 1993.

———. *First Freedom: The Responses of Alabama's Blacks to Emancipation and Reconstruction*. Westport, Conn.: Greenwood Press, 1972.

———. Review of *To Have and to Hold: Slave Work and Family Life in Antebellum South Carolina* by Larry E. Hudson, Jr. *American Historical Review* 102 (Dec. 1997): 1578–79.

———. "Scalawags, Carpetbaggers, and Reconstruction: A Quantitative Look at Southern Congressional Politics, 1868–1872," *Journal of Southern History* 45 (Feb. 1979): 63–76.

———. "The Variable Institution." *Journal of American Ethnic History* 18 (Winter 1999): 111–21.

Lander, Ernest M., Jr. "Charleston: Manufacturing Center of the Old South." *Journal of Southern History* 26, no. 3 (1960): 330–51.

Landes, David S. "French Entrepreneurship and Industrial Growth in the Nineteenth Century," *Journal of Economic History* 9 (1949): 45–61.

———. *The Unbound Prometheus: Technological Change and Industrial Development in Western Europe from 1750 to the Present*. Cambridge: Cambridge University Press, 1969.

Lesser, Charles H. *Relic of the Lost Cause: The Story of South Carolina's Ordinance of Secession*. Columbia: South Carolina Department of Archives and History, 1996.

Lewis, Kenneth E., and Donald L. Hardesty. "Middleton Place: Initial Archaeological Investigations at an Ashley River Rice Plantation." Research Manuscript Series 148. Columbia: Institute of Archaeology and Anthropology, University of South Carolina, 1979.

Linder, Suzanne C. *Historical Atlas of the Rice Plantations of the ACE River Basin 1860*. Columbia: South Carolina Department of Archives and History, 1995.

Litwack, Leon F. *Been in the Storm So Long: The Aftermath of Slavery*. New York: Vintage Books, 1980.

Loy, Wesley. "10 Rumford Place: Doing Confederate Business in Liverpool." *South Carolina Historical Magazine* 98 (Oct. 1997): 349–74.

Malde, Harold E. "Geology of the Charleston Phosphate Area, South Carolina." *U.S. Geological Survey Bulletin* 1079 (1959): 1–105.

Mancini, Matthew J. *One Dies, Get Another: Convict Leasing in the American South, 1866–1928*. Columbia: University of South Carolina, 1996.

Mandle, Jay R. "Black Economic Entrapment After Emancipation in the United States." In *The Meaning of Freedom: Economics, Politics, and Culture After Slavery*, edited by Frank McGlynn and Seymour Drescher, 69–84. Pittsburgh: University of Pittsburgh Press, 1992.

———. *The Roots of Black Poverty: The Southern Plantation Economy After the Civil War*. Durham: Duke University Press, 1978.

Manning, K. R. *Black Apollo of Science: The Life of Ernest Everett Just*. New York: Oxford University Press, 1983.

Marcus, Alan I. "Setting the Standard: Fertilizers, State Chemists, and Early National Commercial Regulation, 1880–87." *Agricultural History* 61 (Winter 1987): 47–73.

Markham, Jesse W. *The Fertilizer Industry: Study of an Imperfect Market*. Nashville: Vanderbilt University Press, 1958.

Maslyn, Williams. *The Phosphateers: A History of the British Phosphate Commissioners and the Christmas Island Phosphate Commission*. Calton, Victoria: Melbourne University Press, 1985.

Mathew, William M. *Agriculture, Geology, and Society in Antebellum South Carolina: The Private Diary of Edmund Ruffin, 1843*. Athens: University of Georgia Press, 1988.

———. *Edmund Ruffin and the Crisis of Slavery in the Old South: The Failure of Agricultural Reform*. Athens: University of Georgia Press, 1988.

McGuire, Mary Jennie. "Getting Their Hands on the Land: The Revolution in St. Helena Parish, 1861–1900." Ph.D. diss., University of South Carolina, 1985.

McPherson, James M. *Battle Cry of Freedom: The Civil War Era*. New York: Ballantine Books, 1989.

Miller, Randall M. "A Yankee Deep in the Heart of Dixie." Review of *The Conquest of Labor: Daniel Pratt and Southern Industrialization*, by Curtis J. Evans. Online posting. Oct. 2002. H-NET Book Review. http://www.h-south@h-net.msu.edu.

Mitchell, Betty L. *Edmund Ruffin: A Biography*. Bloomington: Indiana University Press, 1981.

Mitchell, Margaret. *Gone with the Wind*. New York: Macmillan, 1936.

Moore, Alexander. *Biographical Directory of the South Carolina House of Representatives 1816–1828*. Vol. 5. Columbia: South Carolina Department of Archives and History, 1992.

Moore, Barrington. *Social Origins of Dictatorship and Democracy*. Boston: Beacon Press, 1966.

Moore, James T. "Redeemers Reconsidered: Change and Continuity in the Democratic South, 1870–1900." *Journal of Southern History* 44 (Aug. 1978): 357–78.

Morgan, James Morris. *Recollections of a Rebel Reefer*. Boston: Houghton Mifflin, 1917.

Morgan, Philip D. *Slave Counterpoint: Black Culture in the Eighteenth-Century Chesapeake and Lowcountry*. Chapel Hill: University of North Carolina Press, 1998.

———. "Task and Gang Systems: The Organization of Labor on New World Plantations." In *Work and Labor in Early America*, edited by Stephen Innes, 189–220. Chapel Hill: University of North Carolina Press, 1988.

———. "Work and Culture: The Task System and the World of Lowcountry Blacks, 1700–1880." *William and Mary Quarterly* 39 (Oct. 1982): 563–97.

Murray, Chalmers S. *This Our Land: The Story of the Agricultural Society of South Carolina*. Charleston: Carolina Art Association, 1949.

"Necrology." *South Carolina Historical and Genealogical Magazine* 5 (Jul. 1904): 194–96.

Nepveux, Ethel Trenholm Seabrook. *George Alfred Trenholm and the Company that Went to War 1861–1865*. Anderson: Electric City Printing, 1973.

O'Brien, Gail Williams. *The Legal Fraternity and the Making of a New South Community*. Athens: University of Georgia Press, 1986.

Oakes, James. *Ruling Race: A History of American Slaveholders*. New York: Vintage Books, 1983.

———. *Slavery and Freedom: An Interpretation of the Old South*. New York: Alfred A. Knopf, 1990.

Painter, Carvel. "The Recovery of Confederate Property and Other Assets Abroad, 1865–73." Ph.D. diss., American University, 1973.

Perkins, Edwin J. *Financing Anglo-American Trade: The House of Brown, 1800–1880*. Cambridge: Harvard University Press, 1975.

Phillips, Ulrich Bonnell. *American Negro Slavery: A Survey of the Supply, Employment and Control of Negro Labor As Determined by the Plantation Regime*. 1918. Reprint, Baton Rouge: Louisiana State University Press, 1966.

"Pinckney." *South Carolina Historical and Genealogical Magazine* 39 (Jan. 1938): 25–33.

Poston, Jonathan H. *The Buildings of Charleston: A Guide to the City's Architecture*. Columbia: University of South Carolina Press, 1997.

Powell, Lawrence N. *New Masters: Northern Planters During the Civil War and Reconstruction*. New Haven: Yale University Press, 1980.

Powers, Bernard E., Jr. *Black Charlestonians: A Social History, 1822–1885*. Fayetteville: University of Arkansas Press, 1994.

———. "Community Evolution and Race Relations in Reconstruction Charleston, South Carolina." *South Carolina Historical Magazine* 95 (Jan. 1994): 27–46. Reprint, *South Carolina Historical Magazine* 101 (July 2000): 214–33.

Pruneau, Leigh. Review of *To Have and to Hold: Slave Work and Family Life in Antebellum South Carolina*, by Larry E. Hudson, Jr. *South Carolina Historical Magazine* 99 (Jan. 1998): 95–97.

Prymak, Andrew. Review of *The Fragile Fabric of Union: Cotton, Federal Politics, and the Global Origins of the Civil War*, by Brian Schoen. *South Carolina Historical Magazine* 112 (Jan.–Apr. 2011): 97–99.

Rabinowitz, Howard N. *The First New South: 1865–1920*. Arlington Heights, IL: Harlan Davidson, 1992.

Ransom, Roger L., and Richard Sutch. *One Kind of Freedom: The Economic Consequences of Freedom*. Cambridge: Cambridge University Press, 1977.

Ravenel, Mrs. St. Julien (Harriet Horry Rutledge). *Charleston: The Place and the People*. 1906. Reprint, Easley, S.C.: Southern Historical Press / Strode Publishers, 1972.

Reidy, Joseph P. *From Slavery to Agrarian Capitalism in the Cotton Plantation South: Central Georgia, 1800–1880*. Chapel Hill: University of North Carolina Press, 1992.

Reynolds, Emily B., and John R. Faunt. *Biographical Directory of the Senate of the State of South Carolina 1776–1964*. Columbia: South Carolina Archives Department, 1964.

Reynolds, John S. *Reconstruction in South Carolina 1865–1877*. 1905. Reprint, New York: Negro Universities Press, 1969.

Richards, H. G., and A. H. Hopkins. "Oligocene Fossils from the Old Bolton Phosphate Mine near Charleston, South Carolina." *Geologic Notes* 4, no. 3 (May–June 1960): 19–24.

Richardson, E. B. "Dr. Anthony Cordes and Descendants." *South Carolina Historical and Genealogical Magazine* 43 (Oct. 1942): 219–42.

———. "Dr. Anthony Cordes and Descendants." *South Carolina Historical and Genealogical Magazine* 44 (Jan. 1943): 17–42.

Richardson, Heather Cox. *The Death of Reconstruction: Race, Labor, and Politics in the Post-Civil War North, 1865–1901*. Cambridge: Harvard University Press, 2001.

Ripley, Alexandra. *Scarlett: the Sequel to Margaret Mitchell's Gone with the Wind*. New York: Warner Books, 1991.

Ritter, Gretchen. *Goldbugs and Greenbacks: The Antimonopoly Tradition and the Politics of Finance in America, 1865–1896*. New York: Cambridge University Press, 1997.

Roark, James L. *Masters Without Slaves: Southern Planters in the Civil War and Reconstruction*. New York: W. W. Norton, 1977.

Robinson, Gloria. "Charles Upham Shepard." In *Benjamin Silliman and his Circle: Studies on the Influence of Benjamin Silliman on Science in America*, edited by Leonard G. Wilson, 85–103. New York: Science History Publications, 1979.

Rogers, George C., Jr. *Generations of Lawyers: A History of the South Carolina Bar*. Columbia: South Carolina Bar Foundation, 1992.

Rose, Willie Lee. *Rehearsal for Reconstruction: The Port Royal Experiment*. Indianapolis: Bobbs-Merrill, 1964.

Rosen, Robert N. *The Jewish Confederates*. Columbia: University of South Carolina Press, 2000.

Rosengarten, Theodore. *Tombee: Portrait of a Cotton Planter*. New York: Williams Morrison, 1986.

Rossiter, Margaret W. *Emergence of Agricultural Science: Justus Liebig and the Americans, 1840–1880*. New Haven: Yale University Press, 1975.

Sanders, Albert E. "Additions to the Pleistocene Mammal Faunas of South Carolina, North Carolina, and Georgia." *Transactions of the American Philosophical Society* 92, no. 5 (2002): i–v, 1–152.

———. "Excavation of Oligocene Marine Fossil Beds Near Charleston, South Carolina." *National Geographic Society Research Reports* 12 (1980): 601–21.

———. Unpublished diagram. Charleston Museum. 1999.

Sanders, Albert E., and William D. Anderson Jr. *Natural History Investigations in South Carolina from Colonial Times to the Present*. Columbia: University of South Carolina Press, 1999.

Sanford, Marshall C., Jr. *The Progression of Coosaw Plantation into the 20th Century*. Privately Printed, 1981, Pamphlet, SCHS.

Sass, Herbert Ravenel. "The Rice Coast: Its History and Meaning." In *A Carolina rice plantation of the fifties; 30 paintings in water-colour by Alice R. Huger Smith, narrative by Herbert Ravenel Sass, with chapters from the unpublished memoirs of D. E. Huger Smith*, by Alice R. Huger Smith. New York: Williams Morrow, 1936.

———. "The Story of Little David." *Harper's Magazine* 186 (May 1943): 620–25.

Saville, Julie. *The Work of Reconstruction: From Slave to Wage Laborer in South Carolina, 1860–1870*. New York: Cambridge University Press, 1994.

Schroeder, Glenna R. "'We Will Support the Govt. to the Bitter End': The Augusta Office of the Confederate Nitre and Mining Bureau." *Georgia Historical Quarterly* 70 (Summer 1986): 288–305.

Schwalm, Leslie A. *A Hard Fight for We: Women's Transition from Slavery to Freedom in South Carolina*. Urbana: University of Illinois Press, 1997.

Scott, James C. *Weapons of the Weak: Everyday Forms of Peasant Resistance*. New Haven: Yale University Press, 1985.

Seip, Terry L. *The South Returns to Congress: Men, Economic Measures, and Intersectional Relationships, 1868–1879*. Baton Rouge: Louisiana State University Press, 1983.

Seymour, Liz. "Vive l'heritage Hugeno." *Kiawah Island Legends* (1999): 20–27.

Shapiro, Karin A. *A New South Rebellion: The Battle Against Convict Labor in the Tennessee Coalfields, 1871–1896*. Chapel Hill: University of North Carolina Press, 1998.

Sheridan, Richard C. "Chemical Fertilizers in Southern Agriculture." *Agricultural History* 53 (Jan. 1979): 308–18.

Shick, Tom W., and Don H. Doyle. "The South Carolina Phosphate Boom and the Stillbirth of the New South, 1867–1920." *South Carolina Historical Magazine* 86 (Jan. 1985): 1–31.

Simkins, Francis B., and Robert H. Woody. *South Carolina During Reconstruction*. Chapel Hill: University of North Carolina Press, 1932.

Skaggs, Jimmy M. *The Great Guano Rush: Entrepreneurs and American Overseas Expansion*. New York: St. Martin's Press, 1994.

Smith, H.A.M. "The Ashley River: Its Seats and Settlements." *South Carolina Historical and Genealogical Magazine* 20 (Jan. 1919): 3–51.

Spence, E. Lee. *Treasures of the Confederate Coast: The "Real Rhett Butler" and other Revelations*. Charleston: Narwhal Press, 1995.

Stanley, Amy Dru. *From Bondage to Contract: Wage Labor, Marriage, and the Market in the Age of Slave Emancipation*. Cambridge: Cambridge University Press, 1998.

Stauffer, Michael E. *The Formation of Counties in South Carolina*. Columbia: South Carolina Department of Archives and History, 1994.

Stephens, Lester D. *Ancient Animals and Other Wondrous Things: The Story of Francis Simmons Holmes*. Charleston: Contributions from the Charleston Museum, 1988.

Stevenson, Mary, comp. *The Diary of Clarissa Adger Bowen, Ashtabula Plantation 1865, with excerpts from other Family Diaries and Comments by her Granddaughter, Clarissa Walton Taylor, and Many Accounts of the Pendleton Clemson Area, South Carolina, 1776–1889*. Pendleton, S.C.: Foundation for Historic Preservation in the Pendleton Area, 1973.

Stewart, Mart A. "Rice, Water, and Power: Landscapes of Domination and Resistance in the Lowcountry, 1790–1880." *Environmental History Review* 15 (Fall 1991): 47–64.

Strickland, John Scott. "'No More Mud Work': The Struggle for the Control of Labor and Production in Low Country South Carolina, 1863–1880." In *The Southern Enigma: Essays on Race, Class, and Folk Culture*, edited by Walter J. Fraser Jr. and Winfred B. Moore Jr., 43–62. Westport, Conn.: Greenwood Press, 1983.

———. "Traditional Culture and Moral Economy: Social and Economic Change in the South Carolina Low Country, 1865–1910." In *The Countryside in the Age of Capitalist Transformation: Essays in the Social History of Rural America*, edited by Steven Hahn and Jonathan Prude, 141–78. Chapel Hill: University of North Carolina Press, 1985.

Taylor, David, ed. *South Carolina Naturalists: An Anthology, 1700–1860*. Columbia: University of South Carolina Press, 1998.

Taylor, Rosser H. "Commercial Fertilizers in South Carolina." *South Atlantic Quarterly* 29 (Apr. 1930): 179–89.

———. "Fertilizers and Farming in the Southeast, 1840–1950, Part I 1840–1900." *North Carolina Historical Review* 30 (Jul. 1953): 307–28.

———. "The Sale and Application of Commercial Fertilizers in the South Atlantic States to 1900." *Agricultural History* 21 (Jan. 1947): 46–52.

"The Story of the Fertilizer Industry in Baltimore." *Commercial Fertilizer* (Jun. 1938): 24–36.

Thompson, Henry T. *Ousting the Carpetbagger from South Carolina*. Columbia: R.L. Bryan, 1927.

Tindall, George B. *South Carolina Negroes, 1877–1890*. Columbia: University of South Carolina Press, 1952.

———. *The Persistent Tradition in New South Politics*. Baton Rouge: Louisiana State University Press, 1975.

Tischendorf, Alfred P. "A Note on British Enterprise in South Carolina, 1872–1886." *South Carolina Historical Magazine* 56 (Oct. 1955): 196–99.

Tuten, James H. *Lowcountry Time and Tide: The Fall of the South Carolina Rice Kingdom*. Columbia: University of South Carolina Press, 2010.

University of Virginia Geospatial and Statistical Data Center. *United States Historical Census Data Browser.* Online 1998: University of Virginia, http://fisher.lib.virginia.edu/census/.

Wallace, David D. *The History of South Carolina*. Vol. 3. New York: American Historical Society, 1934.

Warren, Leonard. *Joseph Leidy: The Last Man Who Knew Everything*. New Haven: Yale University Press, 1998.

Wetherington, Mark V. *The New South Comes to Wiregrass Georgia, 1860–1910*. Knoxville: University of Tennessee Press, 1994.

Whitney, Richard A. "The History of Phosphate Mining in Beaufort County, 1870–1914." Paper prepared for the Beaufort County Historical Society, 1989. Beaufort County Library, Beaufort, S.C.

Wickwar, W. H. *300 Years of Development Administration in South Carolina*. Columbia: University of South Carolina, 1970.

Wiebe, Robert H. *The Search for Order 1877–1920*. New York: Hill & Wang, 1967.

Wiener, Jonathan M. *Social Origins of the New South: Alabama, 1860–1885*. Baton Rouge: Louisiana State University Press, 1978.

Wilkins, Mira. "The Free-Standing Company, 1870–1914: An Important Type of British Foreign Direct Investment." *Economic History Review* 41 (1988): 259–82.

Williamson, Joel. *After Slavery: The Negro in South Carolina During Reconstruction, 1861–1877*. Chapel Hill: University of North Carolina Press, 1965.

Wines, Richard A. *Fertilizer in America: From Waste Recycling to Resource Exploitation*. Philadelphia: Temple University Press, 1985.

Winters, Rhett Y. *Washington Caruthers Kerr: The Farmers' Advocate of the 1870's*. Raleigh: School of Agriculture and Life Sciences, North Carolina State of the University of North Carolina at Raleigh, 1964.

Wise, Stephen R. *Lifeline of the Confederacy: Blockade Running During the Civil War.* Columbia: University of South Carolina Press, 1988.

Woodman, Harold. *King Cotton and His Retainers: Financing and Marketing the Cotton Crop of the South, 1800–1925.* 1968. Reprint, Columbia: University of South Carolina Press, 1990.

Woodward, C. Vann. *Origins of the New South 1877–1913.* 1951. Reprint, Baton Rouge: Louisiana State University Press, 1971.

Woody, R. H. "The Labor and Immigration Problem of South Carolina During Reconstruction." *Mississippi Valley Historical Review* 18 (Sept. 1931): 195–212.

Wright, Gavin. *Old South, New South: Revolutions in the Southern Economy Since the Civil War.* New York: Basic Books, 1986.

———. "The Economics and Politics of Slavery and Freedom in the U.S. South." In *The Meaning of Freedom: Economics, Politics, and Culture After Slavery,* edited by Frank McGlynn and Seymour Drescher, 85–111. Pittsburgh: University of Pittsburgh Press, 1992.

Young, Jeffrey R. Review of *Them Dark Days: Slavery in the American Rice Swamps* by William Dusinberre. *Journal of Southern History* 63 (May 1997): 400–401.

Zuczek, Richard. *State of Rebellion: Reconstruction in South Carolina.* Columbia: University of South Carolina Press, 1996.

Index

Page numbers followed by *f* and *m* refer to figures and maps, respectively.

Shepherd W. McKinley is a senior lecturer in history at the University of North Carolina at Charlotte. He is the coauthor of *North Carolina: New Directions for an Old Land*.

NEW PERSPECTIVES ON THE HISTORY OF THE SOUTH
Edited by John David Smith

"In the Country of the Enemy": The Civil War Reports of a Massachusetts Corporal, edited by William C. Harris (1999)

The Wild East: A Biography of the Great Smoky Mountains, by Margaret L. Brown (2000; first paperback edition, 2001)

Crime, Sexual Violence, and Clemency: Florida's Pardon Board and Penal System in the Progressive Era, by Vivien M. L. Miller (2000)

The New South's New Frontier: A Social History of Economic Development in Southwestern North Carolina, by Stephen Wallace Taylor (2001)

Redefining the Color Line: Black Activism in Little Rock, Arkansas, 1940–1970, by John A. Kirk (2002)

The Southern Dream of a Caribbean Empire, 1854–1861, by Robert E. May (2002)

Forging a Common Bond: Labor and Environmental Activism during the BASF Lockout, by Timothy J. Minchin (2003)

Dixie's Daughters: The United Daughters of the Confederacy and the Preservation of Confederate Culture, by Karen L. Cox (2003)

The Other War of 1812: The Patriot War and the American Invasion of Spanish East Florida, by James G. Cusick (2003)

"Lives Full of Struggle and Triumph": Southern Women, Their Institutions, and Their Communities, edited by Bruce L. Clayton and John A. Salmond (2003)

German-Speaking Officers in the United States Colored Troops, 1863–1867, by Martin W. Öfele (2004)

Southern Struggles: The Southern Labor Movement and the Civil Rights Struggle, by John A. Salmond (2004)

Radio and the Struggle for Civil Rights in the South, by Brian Ward (2004; first paperback edition, 2006)

Luther P. Jackson and a Life for Civil Rights, by Michael Dennis (2004)

Southern Ladies, New Women: Race, Region, and Clubwomen in South Carolina, 1890–1930, by Joan Marie Johnson (2004)

Fighting Against the Odds: A Concise History of Southern Labor Since World War II, by Timothy J. Minchin (2004; first paperback edition, 2006)

"Don't Sleep With Stevens!": The J. P. Stevens Campaign and the Struggle to Organize the South, 1963–1980, by Timothy J. Minchin (2005)

"The Ticket to Freedom": The NAACP and the Struggle for Black Political Integration, by Manfred Berg (2005; first paperback edition, 2007)

"War Governor of the South": North Carolina's Zeb Vance in the Confederacy, by Joe A. Mobley (2005)

Planters' Progress: Modernizing Confederate Georgia, by Chad Morgan (2005)

The Officers of the CSS Shenandoah, by Angus Curry (2006)

The Rosenwald Schools of the American South, by Mary S. Hoffschwelle (2006; first paperback edition, 2014)

Honor in Command: Lt. Freeman S. Bowley's Civil War Service in the 30th United States Colored Infantry, edited by Keith Wilson (2006)

A Black Congressman in the Age of Jim Crow: South Carolina's George Washington Murray, by John F. Marszalek (2006)

The Spirit and the Shotgun: Armed Resistance and the Struggle for Civil Rights, by Simon Wendt (2007; first paperback edition, 2010)

Making a New South: Race, Leadership, and Community after the Civil War, edited by Paul A. Cimbala and Barton C. Shaw (2007)

From Rights to Economics: The Ongoing Struggle for Black Equality in the U.S. South, by Timothy J. Minchin (2008)

Slavery on Trial: Race, Class, and Criminal Justice in Antebellum Richmond, Virginia, by James M. Campbell (2008; first paperback edition, 2010)

Welfare and Charity in the Antebellum South, by Timothy James Lockley (2008; first paperback edition, 2009)

T. Thomas Fortune the Afro-American Agitator: A Collection of Writings, 1880–1928, by Shawn Leigh Alexander (2008; first paperback edition, 2010)

Francis Butler Simkins: A Life, by James S. Humphreys (2008)

Black Manhood and Community Building in North Carolina, 1900–1930, by Angela Hornsby-Gutting (2009; first paperback edition, 2010)

Counterfeit Gentlemen: Manhood and Humor in the Old South, by John Mayfield (2009; first paperback edition, 2010)

The Southern Mind Under Union Rule: The Diary of James Rumley, Beaufort, North Carolina, 1862–1865, edited by Judkin Browning (2009; first paperback edition, 2011)

The Quarters and the Fields: Slave Families in the Non-Cotton South, by Damian Alan Pargas (2010; first paperback edition, 2011)

The Door of Hope: Republican Presidents and the First Southern Strategy, 1877–1933, by Edward O. Frantz (2011; first paperback edition, 2012)

Painting Dixie Red: When, Where, Why, and How the South Became Republican, edited by Glenn Feldman (2011; first paperback edition, 2014)

After Freedom Summer: How Race Realigned Mississippi Politics, 1965–1986, by Chris Danielson (2011; first paperback edition, 2013)

Dreams and Nightmares: Martin Luther King Jr., Malcolm X, and the Struggle for Black Equality in America, by Britta Waldschmidt-Nelson (2012)

Hard Labor and Hard Time: Florida's "Sunshine Prison" and Chain Gangs, by Vivien M. L. Miller (2012)

Ain't Scared of Your Jail: Arrest, Imprisonment, and the Civil Rights Movement, by Zoe A. Colley (2013; first paperback edition, 2014)

After Slavery: Race, Labor, and Citizenship in the Reconstruction South, edited by Bruce E. Baker and Brian Kelly (2013; first paperback edition, 2014)

Stinking Stones and Rocks of Gold: Phosphate, Fertilizer, and Industrialization in Postbellum South Carolina, by Shepherd W. McKinley (2014; first paperback edition, 2017)

The Path to the Greater, Freer, Truer World: Southern Civil Rights and Anticolonialism, 1937–1955, by Lindsey R. Swindall (2014)

CPSIA information can be obtained
at www.ICGtesting.com
Printed in the USA
FFOW03n1749071017
40684FF

9 780813 064611